AF595339

E-Textiles 2022: The 4th International Conference on the Challenges, Opportunities, Innovations and Applications in Electronic Textiles

E-Textiles 2022: The 4th International Conference on the Challenges, Opportunities, Innovations and Applications in Electronic Textiles

Editors

Steve Beeby
Kai Yang
Russel Torah
Theodore Hughes-Riley

MDPI • Basel • Beijing • Wuhan • Barcelona • Belgrade • Manchester • Tokyo • Cluj • Tianjin

Editors

Steve Beeby
University of Southampton
Southampton, UK

Kai Yang
University of Southampton
Southampton, UK

Russel Torah
University of Southampton
Southampton, UK

Theodore Hughes-Riley
Nottingham Trent University
Nottingham, UK

Editorial Office
MDPI
St. Alban-Anlage 66
4052 Basel, Switzerland

This is a reprint of Proceedings published online in the open access journal *Engineering Proceedings* (ISSN 2673-4591) (available at: https://www.mdpi.com/2673-4591/30/1).

For citation purposes, cite each article independently as indicated on the article page online and as indicated below:

LastName, A.A.; LastName, B.B.; LastName, C.C. Article Title. *Journal Name* **Year**, *Volume Number*, Page Range.

ISBN 978-3-0365-8362-4 (Hbk)
ISBN 978-3-0365-8363-1 (PDF)

Cover image courtesy of University of Southampton

Contents

About the Editors

Steve Beeby

Prof Steve Beeby holds the prestigious position of the Royal Academy of Engineering Chair in Emerging Technologies on the topic of E-Textile Engineering. His research interests involve the application of flexible electronics, smart printable materials and energy harvesting technologies to electronic textiles (e-textiles). He founded the E-Textiles Network and has over 300 publications and an h-Index of 50 with more than 15,000 citations. He is a co-founder of Perpetuum Ltd, a University spin-out based upon vibration energy harvesting formed in 2004.

Kai Yang

Dr Kai Yang is an Associate Professor in Textile Design and Innovation. She is the Head of Research in Fashion and Textiles. Her research interests are electronic textiles, wearable healthcare, and sustainable textiles. She has been awarded an EPSRC Fellowship and two MRC DPFS grants. She is the co-founder of Smart Fabric Inks, founder of Etexsense, and co-chair of the E-textiles Network.

Russel Torah

Dr Russel Torah graduated with a BEng (hons) in Electronic Engineering and an MSc in Instrumentation and Transducers, both from the University of Southampton. In 2004, Russel obtained a PhD in Electronics from the University of Southampton. Since 2005 he has been a full-time researcher at the University of Southampton where he is currently an Associate Professor. In 2011, Dr Torah co-founded Smart Fabric Inks Ltd, specialising in printed smart fabrics. His research interests are currently focused on smart fabric development but he also has extensive knowledge of energy harvesting, sensors, and transducers. Dr Torah has 145 publications, an h-index of 28 and 2 patents.

Theodore Hughes-Riley

Dr Theodore Hughes-Riley is an Associate Professor in Electronic Textiles, where he acts as a research lead for a number of projects within the Advanced Textiles Research Group (ATRG). His current work focusses on the development of electronically active textile (E-textile) devices. Dr Hughes-Riley has a growing international reputation in the area of E-textiles and has authored over 60 publications, including over 35 peer-reviewed journal articles, primarily in the fields of E-textiles and sensors.

Preface to "E-Textiles 2022: The 4th International Conference on the Challenges, Opportunities, Innovations and Applications in Electronic Textiles"

The UK's E-Textiles Network was established in 2018 with the goal of building the electronic textiles community and acting as a bridge between academia and industry. A key objective of the Network was to establish an international conference on the topic where the latest research and developments could be shared and disseminated. E-Textiles 2022 was the 4th edition of the conference and was held in Nottingham (UK) as a hybrid virtual and in-person event. It included a mix of invited and accepted speakers from around the world. Accepted papers were peer-reviewed, and this collection of papers from the conference presents the latest work on a range of electronic textile applications and technologies. The Editors of the collection were responsible for organizing the conference with the assistance of the E-Textiles International Conference Committee and the Technical Programme Committee.

Steve Beeby, Kai Yang, Russel Torah, and Theodore Hughes-Riley
Editors

Editorial

Statement of Peer Review †

Russel Torah [1,*], Kai Yang [2], Theodore Hughes-Riley [3] and Stephen Beeby [1]

[1] School of Electronics & Computer Science, University of Southampton, Southampton SO17 1BJ, UK; spb@ecs.soton.ac.uk
[2] Winchester School of Art, University of Southampton, Southampton SO23 8DL, UK; ky2e09@soton.ac.uk
[3] Advanced Textiles Research Group, Nottingham Trent University, Nottingham NG11 8NS, UK; theo.hughes-riley@ntu.ac.uk
* Correspondence: rnt@ecs.soton.ac.uk

† All Papers Published in the Volume Are Presented at the 4th International Conference on the Challenges, Opportunities, Innovations and Applications in Electronic Textiles, Nottingham, UK, 8–10 November 2022.

In submitting conference proceedings to *Engineering Proceedings*, the volume editors of the proceedings certify to the publisher that all papers published in this volume have been subjected to peer review administered by the volume editors. Reviews were conducted by expert referees to the professional and scientific standards expected of a proceedings journal.

- Type of peer review: triple open reviews;
- Conference submission management system: Web form + Excel;
- Number of submissions sent for review: 56;
- Number of submissions accepted: 39 (12 oral, 27 posters);
- Acceptance rate (number of submissions accepted/number of submissions received): 70% overall, 31% oral;
- Average number of reviews per paper: 3;
- Total number of reviewers involved: 28;
- Any additional information on the review process: None.

Conflicts of Interest: The authors declare no conflict of interest.

Citation: Torah, R.; Yang, K.; Hughes-Riley, T.; Beeby, S. Statement of Peer Review. *Eng. Proc.* **2023**, *30*, 21. https://doi.org/10.3390/engproc2023030021

Published: 29 June 2023

Editorial

Preface of the 4th International Conference on the Challenges, Opportunities, Innovations and Applications in Electronic Textiles (E-Textiles 2022) †

Steve Beeby [1], Kai Yang [2], Russel Torah [1,*], Beckie Isaia [1] and Theodore Hughes-Riley [3]

1 School of Electronics & Computer Science, University of Southampton, Southampton SO17 1BJ, UK; spb@ecs.soton.ac.uk (S.B.); r.j.isaia@soton.ac.uk (B.I.)
2 Winchester School of Art, University of Southampton, Southampton SO23 8DL, UK; ky2e09@soton.ac.uk
3 Advanced Textiles Research Group, Nottingham Trent University, Nottingham NG1 4GG, UK; theo.hughes-riley@ntu.ac.uk
* Correspondence: rnt@ecs.soton.ac.uk
† All papers published in the volume are presented at the 4th International Conference on the Challenges, Opportunities, Innovations and Applications in Electronic Textiles, Nottingham, UK, 8–10 November 2022.

Abstract: The 4th International Conference on the Challenges, Opportunities, Innovations and Applications in Electronic Textiles (E-Textiles 2022) was held in Nottingham (United Kingdom) on 8–10 November 2022.

Keywords: electronics; wearables; smart fabrics; electronic textiles; E-textiles; sensing; healthcare devices; textile antennas; energy harvesting; printed electronics; knitted E-textiles; functional yarns

1. Introduction

The United Kingdom's E-Textiles Network was established in 2018 with the goal of building the electronic textiles community and acting as a bridge between academia and industry. A key objective of the Network was to establish an international conference on the topic where the latest research and developments could be shared and disseminated. The first E-Textiles conference was held in London in 2019, and E-Textiles 2022, the 4th edition of the conference, was held in Nottingham as a hybrid virtual and in-person event. It included a mix of invited and accepted speakers from around the world. Accepted papers were peer-reviewed, and this collection of papers from the conference presents the latest work on a range of electronic textile applications and technologies.

Citation: Beeby, S.; Yang, K.; Torah, R.; Isaia, B.; Hughes-Riley, T. Preface of the 4th International Conference on the Challenges, Opportunities, Innovations and Applications in Electronic Textiles (E-Textiles 2022). *Eng. Proc.* **2023**, *30*, 22. https://doi.org/10.3390/engproc2023030022

Published: 11 July 2023

2. Conference Days

The event was held in Nottingham from the 8th to the 10th of November 2022. The technical program was organized into six sessions covering the following topics: sensing and embedded systems, applications and future trends, manufacturing and standards, reliability and sustainability, design and fashion, and immersive technologies.

The first day of the conference focused on E-textiles in action, where participants could bring demonstrations to showcase, and finished with a talk on E-textiles research at Nottingham Trent University.

The next two days consisted of a mixture of in-person and remote presentations, as well as a panel discussion on "Applications and Future Trends for E-Textiles", and a poster session.

There were two keynote talks, the first being on "The Challenges of Bringing Garment Based E-Textiles to the Mass Consumer Market" by Martin Ashby (Prevayl, UK). The second keynote talk, "Smart Textiles for Personalized Health Care", was presented by Dr Jun Chen (UCLA, USA).

Fifty-six papers were submitted to the conference, and twenty are published in a volume of MDPI's *Engineering Proceedings* [1].

3. E-Textiles 2022 Prizes

E-Textiles 2022 awarded three prizes, which were sponsored by MDPI *Micromachines*.

3.1. Best Paper Award

Winner: Michelle Farrington (AFFOA, USA) "New Frontiers in Advanced Fibers and Fabrics".

3.2. Best Student Paper Award

Winner: Xiaona Yang (Donghua University, PRC) "Direct Wet-Spun Single-Walled Carbon Nanotubes-Based p-n Segmented Filaments toward Wearable Thermoelectric Textiles".

3.3. Best Student Poster Award

Winner: Chengning Yao (Imperial College London, UK) "Thermally conductive h-BN/polymer composites for e-textiles thermal management".

4. Committees

Conference chair
Dr Theodore Hughes Riley, Nottingham Trent University
Local Organizing Committee
Prof. Tilak Dias, Nottingham Trent University
Dr Yang Wei, Nottingham Trent University
Dr Arash Shahidi, Nottingham Trent University
Kalana Marasinghe, Nottingham Trent University
Dr Pasindu Lugoda, Nottingham Trent University
Jasbir Shanker, Defence and Security Accelerator (DASA)
Professor Steve Morgan, University of Nottingham
Technical Programme Committee
Dr Russel Torah (Chair), University of Southampton
Dr Jun Chen, University of California, Los Angeles (UCLA)
Dr Jess Jur, North Carolina State University
Prof. Anne Schwarz-Pfeiffer, Hochschule Niederrhein
Sigrid Rotzler, Fraunhofer IZM
Mark Catchpole, Conductive Transfers
Dr Yang Wei, Nottingham Trent University
Dr Yi Li, University of Southampton
Dr Ishara Dharmasena, Loughborough University
Dr Rebecca Stewart, Imperial College London
Dr Debra Carr, Defence and Security Accelerator (DASA)
Dr Sasikumar Arumugam, University of Southampton
Jessica Saunders, London College of Fashion
Dr Abiodun Komolafe, University of Southampton
Dr Theodore Hughes-Riley, Nottingham Trent University
Dr Arash Moghaddassian Shahidi, Nottingham Trent University
Jessica Stanley, Nottingham Trent University
Dr Sheng Yong, University of Southampton
Prof. John Wilson, Power Textiles Limited
Dr Kun Zhang, Donghua University
Dr Kai Yang, University of Southampton
Jane McCann, University of Wales, Newport
International Conference Steering Committee

Prof. Steve Beeby (Chair), University of Southampton
Dr Kai Yang, University of Southampton
Dr Jess Jur, North Carolina State University
Christian Dalsgaard, Smart Textiles Alliance
Malte von Krshiwoblozki, Fraunhofer IZM
Prof. Vladan Koncar, GEMTEX Research Laboratory, ENSAIT
Prof. George Stylios, Heriot-Watt University
Dr Laura Salisbury, Royal College of Art
Dr Kun Zhang, Donghua University
Prof. Anne Schwarz-Pfeiffer, Hochschule Niederrhein
Dr Sara Robertson, Royal College of Art
Prof. Henry Yi Li, University of Manchester
Dr Jun Chen, University of California, Los Angeles (UCLA)
Techincal Support
Beckie Isaia, University of Southampton

5. Sponsorship

E-Textiles 2022 would like to thank the following sponsors for supporting this event:
MDPI *Micromachines*
Softmatter
Medical Technologies Innovation Facility (MTIF)

It is our great pleasure to announce that the 5th International Conference on the Challenges, Opportunities, Innovations and Applications in Electronic Textiles (E-Textiles 2023) will be held in Ghent, Belgium from the 14th to the 16th of November 2023 (https://e-textilesconference.com/).

Conflicts of Interest: The authors declare no conflict of interest.

Reference

1. E-Textiles 2022 Conference Proceedings. Available online: https://www.mdpi.com/2673-4591/30/1 (accessed on 21 June 2023).

Proceeding Paper

Optimization of Heating Performance of the Rib-Knitted Wearable Heating Pad †

Sandeep Kumar Maurya, Apurba Das and Bipin Kumar *

Department of Textile and Fibre Engineering, Indian Institute of Technology Delhi, New Delhi 110016, India
* Correspondence: bipin.kumar@textile.iitd.ac.in

† Presented at the 4th International Conference on the Challenges, Opportunities, Innovations and Applications in Electronic Textiles, Nottingham, UK, 8–10 November 2022.

Abstract: Textile structures such as knitting, weaving, braiding, and nonwoven can be used to produce wearable heating pads. Knitted fabric has unique properties such as stretchability, flexibility, and comfort among these textile structures. However, traditional knitted heating pads are manufactured by employing a straightforward three-structure design comprised of a plain, rib, and interlock with yarn that is entirely conductive. The usage of fully conductive materials in industrial applications has been restricted primarily as a result of their more expensive price tag and higher power requirements. Herein, we reported a rib (knitting structure)-based wearable heater with localized conductive yarn. A 14-gauge V-bed knitting machine is used to prepare a localized heating pad with a slight variation in the loop length. The rib-knitted structure (R1) with the lowest loop length showed a 47.4 °C average surface temperature at 9 Volt DC power source. The laboratory-based prototype of the heating pad is also designed for alleviation of joint and muscle pain in the affected area of the body.

Keywords: smart textiles; wearable heating pad; knitted structures; heat therapy; conductive textiles

Citation: Maurya, S.K.; Das, A.; Kumar, B. Optimization of Heating Performance of the Rib-Knitted Wearable Heating Pad. *Eng. Proc.* **2023**, *30*, 1. https://doi.org/10.3390/engproc2023030001

Academic Editors: Steve Beeby, Kai Yang, Russel Torah and Theodore Hughes-Riley

Published: 19 January 2023

1. Introduction

Nowadays, conductive yarn or fabrics are utilized for smart wearable electronic textiles. Textile-based wearable heaters have drawn attention to the application of heat therapy for the reliving of joint and muscle pain. A higher skin temperature increases the heat circulation and blood flow in a specific area of our body. Heating devices can be divided into four categories: electrical-, phase change material-, chemical-, and fluid-based heating [1]. Electrical heating is the most preferred for wearable applications since it requires less time to reach an equilibrium temperature. Therefore, many researchers focused on textile-based electrical heating products such as gloves, belts, and pads for heat therapy applications [2–4]. Equation (1) can be used to calculate the amount of heat that is produced when an electric current flows through a conductor.

$$Heat\ dissipated\ (H) = I^2 \times R \quad (1)$$

where I and R are the current flowing in the conductor and the resistance of the conductor, respectively.

Recently, textile-based wearable heaters have gained a lot of attention owing to their intrinsically flexible, soft, breathable properties and efficient production techniques. Various techniques for incorporating heating components into woven [5,6], knitted [7,8], nonwoven [9], and embroidered [10] fabrics have been found in the literature. The woven structure lacks flexibility due to the horizontal and vertical interlacement of yarns, limiting the yarn mobility in the structure. Heating fabric made up of nonwoven fabric has a limited use owing to the high electrical resistance of conductive fabric. The knitted-based heating fabric has a significant advantage over other methods in terms of its flexibility, stretchability, and comfortability. Liu et al. [11] designed and fabricated the three types of knitted heating fabrics (KHFs), plain, rib, and interlock fabric, using silver-plated conductive yarn and polyester spun yarn. They concluded that the interlock structure was superior to the plain and rib-knitted structures in terms of the heating performance. The structural elements of the knitted structure play an important role for optimizing the heating performance. Kexia sun et al. [12] reported that float and tuck stiches decrease the resistance value of conductive knitted fabric. Consequently, the surface temperature of the combination of a float and tuck knitted heating pad was higher than that of the 100% knitted structure.

The literature has given very little justification on the optimization of the heating performance of the knitted-based heating pad. Herein, we reported a rib (knitting structure)-based wearable heater with localized conductive yarn. This study investigated the surface temperature of the heating pad with a slight variation in the loop length in the knitted structure.

2. Materials and Methods

Low twisted cotton yarn (74 tex) and silver coated nylon yarn (30 tex) were used to fabricate the heating pad by using a V-bed knitting machine (14 gauge). Cotton yarn was used as a nonconductive material, while silver coated yarn was used as a conductive material. The rib-knitted structure was prepared with three courses of localized conductive yarn after six courses of non-conductive yarn, as shown in Figure 1a–e. Four samples of heating pads, such as R1, R2, R3, and R4, were prepared with a varying loop length. The specification of the samples is illustrated in Table 1.

Table 1. Sample specifications.

Sample Code	Sample Appearance	Electrical Resistance (Ω)	Areal Density (g/m²)	Loop Length (mm)	Courses Per Inch (CPI)	Wales Per Inches (WPI)	Stich Density (loops/inch²)
R1		51 ± 2	414 ± 10	4.82	27	33	891
R2		63 ± 2	346 ± 10	5.12	22	30	660
R3		75 ± 2	330 ± 10	5.81	18	28	504
R4		80 ± 2	309 ± 10	6.21	16	26	416

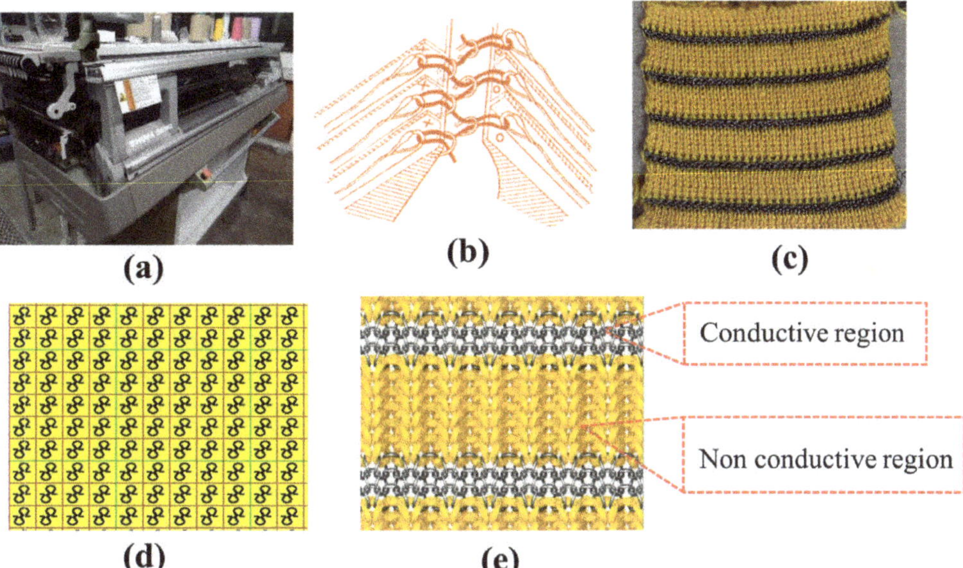

Figure 1. Fabrication of knitted structures (heating pad), (**a**) V-bed knitting machine, (**b**) needles arrangement on V-bed, (**c**) actual image of the heating pad, (**d**) computerized design for rib structure, (**e**) mimetic view of the samples.

3. Results and Discussion

Analysis of Surface Temperature of Knitted Structures

The average surface temperature of the rib-knitted heating pad was analyzed by a thermal infrared camera (Fluke ti450 pro) with an under specified voltage source. The DC source (IT6720-100W) was used to power the heating pad's active component. The sample was kept at a distance of 40 cm from the camera, which was fixed on a tripod with 95% emissivity.

The surface temperature of the fabric was investigated at a 9-volt DC power source at room temperature. The rib structure (R4) with a loop length of 6.21 mm showed a 31.4 °C average surface temperature. At the same input power source, the heating performance of the rib structure increased by 51% with a loop length of 4.82 mm, as shown in Figure 2a. This can be explained by the tightness factor of the knitted structures. The tightness factor of the knitted fabric is calculated by the following Equation (2) [13].

$$Tightness\ factor\ (TF) = \frac{\sqrt{Tex}}{l} \tag{2}$$

where l is the loop length of the rib structure in cm and Tex is the linear density of the yarn.

The fabric sample (R1) with the shorter loop length exhibited a more tightness factor, i.e., 11.36 (Table 2), resulting in a more compact and less conductive yarn being required on the active part of the sample. Thus, knitted structures (R1) exhibited a lower electrical resistance, i.e., 51 Ω, than the other samples. Consequently, the amount of current flowing through the knitted structures (R1) increases, leading to an increase in the fabric's surface temperature. A sample (R1) is utilized to design the heating pad to test the viability of the knitted structure. The heating pad is applied to the elbow, wrist, and calf regions of the body part at a 9 V/2.5 A DC power source, as shown in Figure 2b–d.

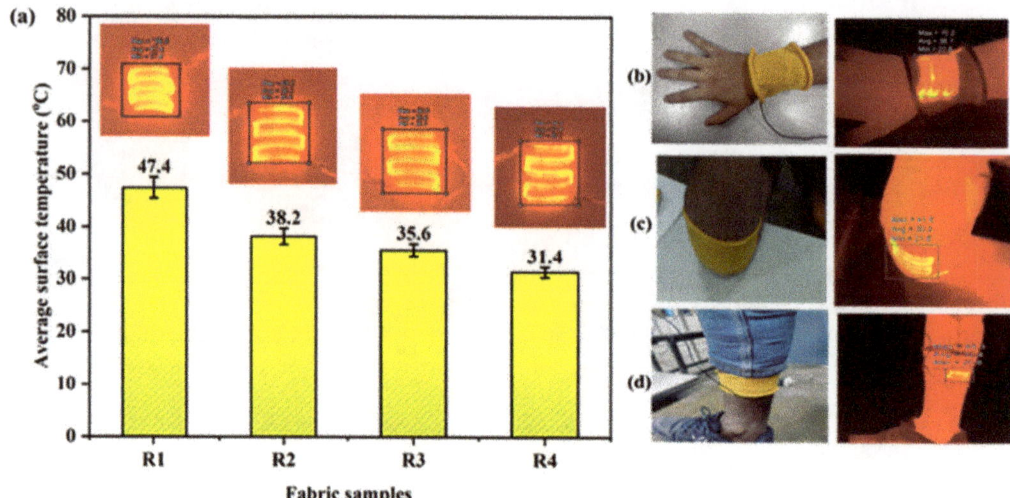

Figure 2. (**a**) Result of the surface temperature of knitted structures at 9 volts and knitted-based localized heating pad developed in the lab for joint and muscle pain in the (**b**) wrist, (**c**) elbow, and (**d**) calf areas.

Table 2. Tightness factor of the rib-knitted structures.

Sample Code	R1	R2	R3	R4
Tightness factor	11.36	10.69	9.43	8.82

4. Conclusions

The objective of this study was to optimize the surface temperature of the heating pad by varying the loop length of the knitted structure, a subtle alteration in the structure of the loop that has a significant impact on the surface temperature of the heating pad. The sample (R1) with the lowest loop length with a compact structure showed a good heating performance. The prototype of the heating pad is designed for the alleviation of joint and muscle pain in the affected area of the body. The study's findings indicate that the loop length of the knitted fabric is a critical component in optimizing the surface temperature.

Author Contributions: Conceptualization, methodology, validation, investigation, and writing—original draft preparation, S.K.M.; writing—review and editing and supervision, B.K. and A.D. All authors have read and agreed to the published version of the manuscript.

Funding: The author would like to acknowledge financial support from the Indian Institute of Technology Delhi.

Informed Consent Statement: Not applicable.

Data Availability Statement: Data is contained within the article.

Conflicts of Interest: The authors declare no conflict of interest.

References

1. Wang, F.; Gao, C.; Kuklane, K.; Holmér, I. A review of the technology of personal heating garments. *Int. J. Occup. Saf. Ergon.* **2010**, *16*, 387–404. [CrossRef] [PubMed]
2. Hamdani, S.T.A.; Fernando, A.; Maqsood, M. Thermo-mechanical behavior of stainless steel knitted structures. *Heat Mass Transf.* **2016**, *52*, 1861–1870. [CrossRef]
3. Kumar, B.; Das, A.; Sharma, A.; Krishnasamy, J.; Alagirusamy, R. Cyclic bursting loading on needle-punched nonwovens: Part I–Distention behavior. *Indian J. Fibre Text. Res. (IJFTR)* **2018**, *43*, 20–24.

4. Pragya, A.; Singh, H.; Kumar, B.; Gupta, H.; Shankar, P. Designing and investigation of braided-cum-woven structure for wearable heating textile. *Eng. Res. Express* **2020**, *2*, 015003. [CrossRef]
5. Sezgin, H.; Bahadır, S.K.; Boke, Y.; Kalaoğlu, F. Investigation of Heating Behaviour of E-textile Structures. *Int. J. Mater. Text. Eng.* **2015**, *9*, 491–494.
6. Muthukumar, N.; Thilagavathi, G.; Kannaian, T.; Periyasamy, S. Development and Characterization of Metal Woven Electric Heating Fabrics. In *Functional Textiles and Clothing*; Springer: Berlin/Heidelberg, Germany, 2019; pp. 119–127.
7. Hao, L.; Yi, Z.; Li, C.; Li, X.; Yuxiu, W.; Yan, G. Development and characterization of flexible heating fabric based on conductive filaments. *Measurement* **2012**, *45*, 1855–1865. [CrossRef]
8. Liu, S.; Yang, C.; Zhao, Y.; Tao, X.M.; Tong, J.; Li, L. The impact of float stitches on the resistance of conductive knitted structures. *Text. Res. J.* **2016**, *86*, 1455–1473. [CrossRef]
9. Xu, J.; Xin, B.; Du, X.; Wang, C.; Chen, Z.; Zheng, Y.; Zhou, M. Flexible, portable and heatable non-woven fabric with directional moisture transport functions and ultra-fast evaporation. *RSC Adv.* **2020**, *10*, 27512–27522. [CrossRef] [PubMed]
10. Petcu, I.; Agrawal, P.; Curteza, A.; Brinks, G.; Teodorescu, M. A comparative study of heating elements used for the development of textile heaters. In Proceedings of the 12th AUTEX World Textile Conference, Faculty of Textile Technology of the University of Zagreb, Zadar, Croatia, 13–15 June 2012.
11. Liu, H.; Li, J.; Chen, L.; Liu, L.; Li, Y.; Li, X.; Li, X.; Yang, H. Thermal-electronic behaviors investigation of knitted heating fabrics based on silver plating compound yarns. *Text. Res. J.* **2016**, *86*, 1398–1412. [CrossRef]
12. Sun, K.; Liu, S.; Long, H. Structural parameters affecting electrothermal properties of woolen knitted fabrics integrated with silver-coated yarns. *Polymers* **2019**, *11*, 1709. [CrossRef] [PubMed]
13. Maurya, S.K.; Uttamrao Somkuwar, V.; Garg, H.; Das, A.; Kumar, B. Thermal protective performance of single-layer rib-knitted structure and its derivatives under radiant heat flux. *J. Ind. Text.* **2021**, *51*, 8865S–8883S. [CrossRef]

Proceeding Paper

A Screen-Printed 8 × 8 Pixel Electroluminescent Display on Fabric †

Huanghao Dai *, Thomas Greig, Russel Torah and Steve Beeby

School of Electronics & Computer Science, University of Southampton, Southampton SO17 1BJ, UK
* Correspondence: hd2m19@soton.ac.uk
† Presented at the 4th International Conference on the Challenges, Opportunities, Innovations and Applications in Electronic Textiles, Nottingham, UK, 8–10 November 2022.

Abstract: A screen-printed electroluminescent (EL) matrix display on fabric is presented. This work demonstrates the fabrication of a screen-printed multilayer display with a matrix of 8 × 8 pixels, as well as the design and construction of integrated drive electronics capable of operating the EL display and achieving good visibility. Each pixel is 1 mm × 1 mm, which is smaller than in previously reported literature. The EL matrix was successfully printed and laminated on fabric at this higher resolution, improving the visual effect and decreasing the overall display size, and reducing the impact on the flexibility and breathability of the underlying fabric. This proof-of-concept demonstrator EL display shows the potential for more complex pixel displays for e-textile and printed electronic applications, such as interactive clothing, and information displays in applications, such as the automotive field, architecture, and point of sale advertising.

Keywords: electroluminescent devices; e-textiles; screen printing; smart fabrics; flexible electronics

Citation: Dai, H.; Greig, T.; Torah, R.; Beeby, S. A Screen-Printed 8 × 8 Pixel Electroluminescent Display on Fabric. *Eng. Proc.* **2023**, *30*, 2. https://doi.org/10.3390/engproc2023030002

Academic Editors: Kai Yang and Theodore Hughes-Riley

Published: 19 January 2023

1. Introduction

Printed electroluminescence (EL) materials are widely used for the fabrication of long-lasting, high-intensity light-emitting devices on flexible substrates [1]. The main benefits are the ease of fabrication of an EL multilayer structure via screen printing, which has a higher layer-thickness tolerance compared with equivalent printed OLEDs. In addition, the polymer inks required for fabrication are more widely available. The previous literature has successfully implemented an EL display on fabrics to fabricate a watch [2] and a traffic signal system on a backpack [1]. To achieve improved high-resolution display performance, the pixel pitch, which is the distance from the centre of a pixel to the centre of the adjacent pixel, is a critical dimension. A lower pixel pitch results in an image with smoother edges, finer detail, and smaller dimensions in the flexible display, increasing the comfort of wearable e-textile applications.

2. Materials and Methods

2.1. Electroluminescent (EL) Lamp Structure

A standard parallel plate capacitor configuration is used to make EL lamps, with a light-emitting phosphor layer sandwiched between the two electrodes. To allow light to be emitted, one of the electrodes is transparent. Because the phosphor layer would electrically break down at high field strengths, a dielectric layer is also placed between the electrodes to prevent a short circuit. Figure 1 shows the cross-section structure of a complete EL lamp with the materials used for each layer, which consists of four layers [2]. A bus-bar layer is added between the phosphor and the translucent conductor (PEDOT) layer for improved electric field distribution in this design to create an 8 × 8 pixels array.

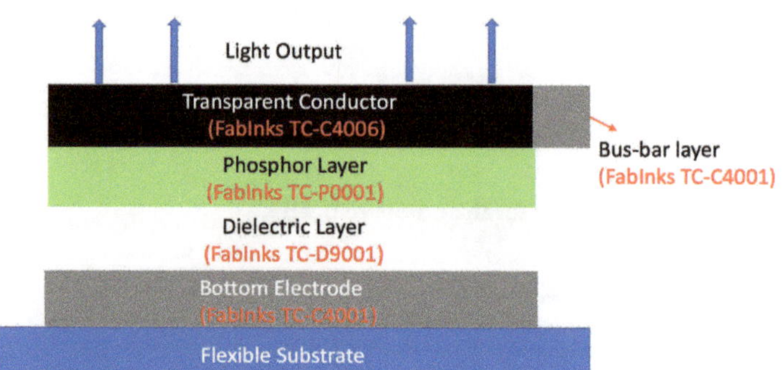

Figure 1. Schematic diagram of the cross-section structure of printed EL lamp with inks used from Smart Fabric Inks Ltd. for each layer, which are labelled in red.

2.2. EL Pixel Arrays Fabrication

The screen-printing method was used to print the flexible EL lamps array with high printing resolution and low cost compared with the clean room facilities typically required for OLED production.

2.2.1. Screen Design

The first step of printing EL lamps is manufacturing the screen. In this project, to test the resolution limitation of screen-printing method, two layouts for the displays were designed using Tanner L-Edit software. The screen design is shown in Figure 2A, which has a print area of 10 cm × 10 cm. It consists of seven large pixel arrays (one of which is labelled with the red shaded area in Figure 2A) and eight combinations of nine small pixel arrays (one of which is labelled with the blue shaded area in Figure 2A), plus six alignment patterns to aid registration of the various layers in the printing process. The screen details (mesh size and emulsion thickness) are given below for each layer.

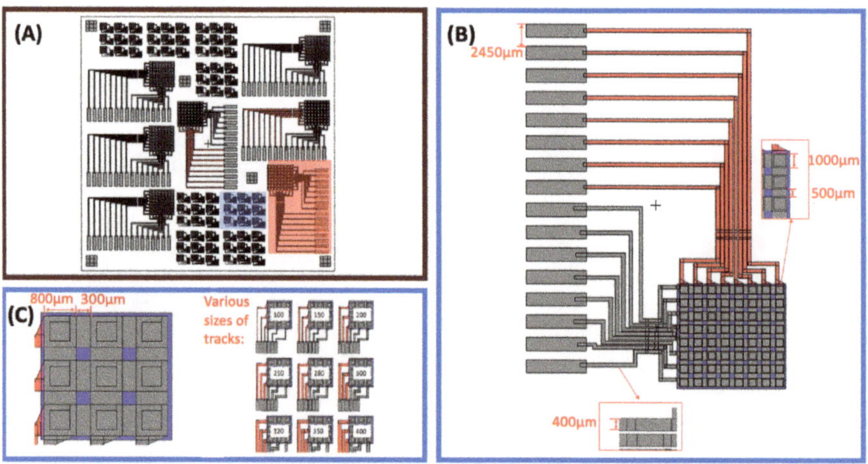

Figure 2. Screen design, which consists of overall layout (**A**), larger 8 × 8 pixels arrays, (**B**) and smaller 3 × 3 pixels arrays (**C**).

In the large pixel 8 × 8 array, the size of each pixel is 1000 μm × 1000 μm with a gap of 500 μm, which is shown in Figure 2B. The linewidth of the connecting tracks is 500 μm, while the width of the connector pitch is 2540 μm to match the flexible printed circuit (FPC) connector used for testing.

The smaller pixel arrays were designed to test the resolution limits of the screen printing. The size of each pixel in the 3 × 3 pixels arrays is 800 µm × 800 µm, with a gap of 300 µm (Figure 2C, left part). The sizes of the connecting tracks of nine different arrays were designed to vary from 100 microns to 400 microns, which is shown in Figure 2C (right part), to explore the printing resolution limitation of screen-printing method.

2.2.2. Screen Printing Process

The EL lamps array was fabricated using five functional layers as detailed below. The schematic pictures of the patterns after printing each layer are shown in Figure 3. A 120-34 polyester screen with 5 µm emulsion thickness was used for each screen except for the busbar, which had 30 µm emulsion to aid the print step across the layers, and the PEDOT layer, which used a 140-30 mesh, identified as the optimum mesh by the ink manufacturer.

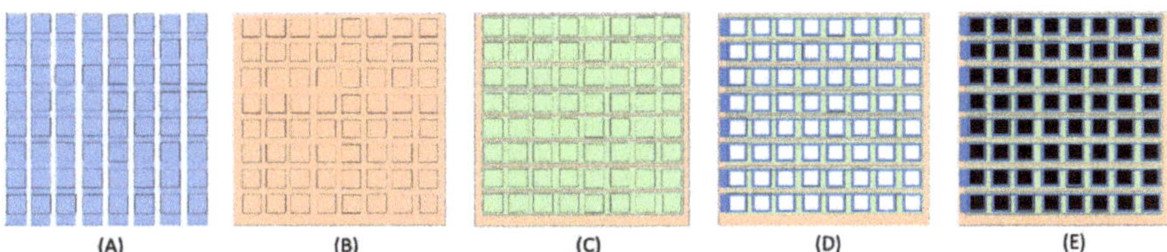

Figure 3. Top views of the EL lamps after the printing of each layer: (**A**) silver layer, (**B**) dielectric layer, (**C**) phosphor layer, (**D**) bus-bar layer, (**E**) PEDOT layer.

The printing process used to produce the EL lamp is:

1. Print silver layer to create the conductive tracks and conductive pads. The paste was cured in a box oven at 120 °C for 10 min;
2. Print dielectric layer to prevent short circuit across the capacitor structure and acts as a light reflector. The paste was cured at 130 °C for 15 min;
3. Print phosphor light-emitting layer. The paste was cured at 130 °C for 15 min;
4. Print busbar layer to create conductive tracks and conductive pads. The paste was cured at 120 °C for 15 min;
5. Print PEDOT layer to allow light emission from the phosphor layer. The paste was cured at 130 °C for 15 min.

3. Results and Discussions

The EL pixels arrays were printed on 120 µm Policrom film and then laminated on to the fabric to achieve the flexible EL display screen on the fabric. The lamps were tested using a standard EL driving circuit, shown in Figure 4, which shows the visibility of each pixel and the retained flexibility of the fabric.

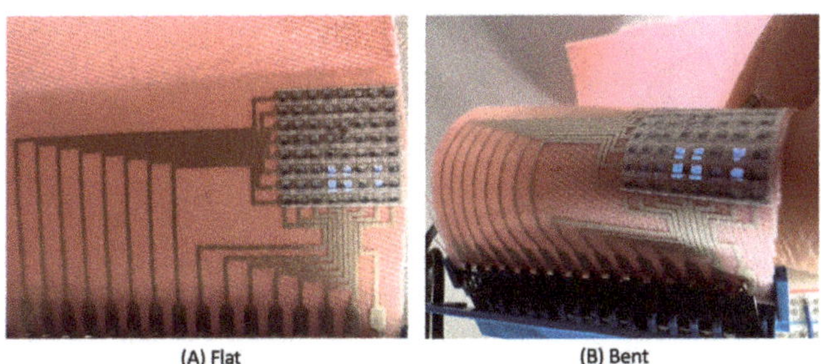

Figure 4. Powering the larger EL pixels arrays in flat (**A**) and bent (**B**) conditions.

The optical properties of the blue emitter were also assessed, which is shown in Figure 5A. The emission peak is at a wavelength of ~550 nm, with a brightness of

~200 cd/m^2. The detected colour of emitting light was shown in CIE 1931 colour space in Figure 5B, of which the coordinates are (0.156, 0.426).

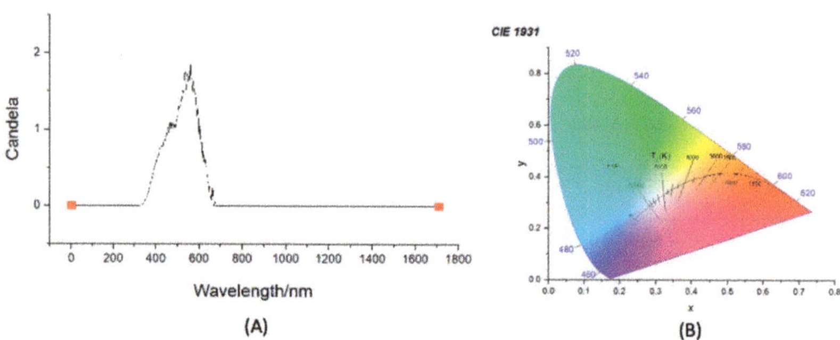

Figure 5. Measured luminance (**A**) and CIE 1931 colour space (blue circle) (**B**).

The smaller 3 × 3 pixels arrays were also tested, and the arrays of sizes greater than or equal to 150 µm were successful. There were defects in the printed 100-microns tracks, which caused open circuits, meaning the pixels could not be lit up. The printed performance of nine different sizes of tracks indicates that the screen-printing method is suitable for fabricating devices with track widths of more than 150 µm.

4. Conclusions

This work improved the screen-printed resolution of electroluminescent (EL) lamps on fabric and is capable of reducing the size of displays whilst retaining the flexible property of fabrics, which helps in reducing the size limit of wearable display devices. However, the limitation is that the screen-printing resolution must be 150 µm to achieve operational displays. Thus, future work must investigate an alternative printing method, namely reverse-offset printing, and use it to fabricate EL lamps to reduce their size by a factor of at least two whilst maintaining good performance.

Author Contributions: H.D. carried out the experiment design and implementation and wrote the manuscript draft. T.G. provided technical support in building the driving circuit. R.T. provided supervision for the research work, screen design, material selection, and electronics and edited the manuscript draft. S.B. supervised the research and edited the manuscript draft. All authors have read and agreed to the published version of the manuscript.

Funding: This work was supported by EPSRC grant Wearable and Autonomous Computing for Future Smart Cities: A Platform Grant (EP/P010164/1).

Institutional Review Board Statement: Not applicable.

Informed Consent Statement: Not applicable.

Data Availability Statement: No new data were created or analyzed in this study. Data sharing is not applicable to this article.

Conflicts of Interest: The authors declare no conflict of interest.

References

1. Ivanov, A. A printed electroluminescent matrix display: Implementation details and technical solutions. In Proceedings of the 2018 IMAPS Nordic Conference on Microelectronics Packaging (NordPac), Oulu, Finland, 12–14 June 2018. [CrossRef]
2. de Vos, M.; Torah, R.; Glanc-Gostkiewicz, M.; Tudor, J. A Complex Multilayer Screen-Printed Electroluminescent Watch Display on Fabric. *J. Disp. Technol.* **2016**, 12, 1757–1763. [CrossRef]

Proceeding Paper

Measuring the Flex Life of Conductive Yarns in Narrow Fabric †

Paula Veske *, Frederick Bossuyt, Filip Thielemans and Jan Vanfleteren

Centre for Microsystems Technology (CMST), Interuniversity Microelectronics Centre (IMEC), Ghent University, 9000 Ghent, Belgium

* Correspondence: paula.veske@ugent.be

† Presented at the 4th International Conference on the Challenges, Opportunities, Innovations and Applications in Electronic Textiles, Nottingham, UK, 8–10 November 2022.

Abstract: The current work presents a testing machine built from off-the-shelf components to test for conductive yarns' (or textiles') durability to repeated bending that can occur during general wear-and-tear or domestic washing procedures. The testing method is explained with an example and results, comparing two different conductive yarns weaved into polyester-based narrow fabric.

Keywords: wearables; e-textiles; electronics; testing; reliability

Citation: Veske, P.; Bossuyt, F.; Thielemans, F.; Vanfleteren, J. Measuring the Flex Life of Conductive Yarns in Narrow Fabric. *Eng. Proc.* **2023**, *30*, 3. https://doi.org/10.3390/engproc2023030003

Academic Editors: Steve Beeby, Kai Yang, Russel Torah and Theodore Hughes-Riley

Published: 19 January 2023

1. Introduction

Conductive yarns are an essential part of e-textile (electronic textiles) applications [1,2]. Nevertheless, integrated electronics often create high-stress areas at the electronics encapsulation–textiles transition points. Thus, measuring their ability to withstand repeated bending is essential. Contrary to domestic washing tests, bending tests help to identify the exact moment(s) when the conductive yarns are breaking, thus helping to compare their durability more accurately and in uniform conditions.

This work presents a bending tester that can be assembled from off-the-shelf components together with a testing method to test e-textiles interconnection materials' durability. The study exemplifying the method includes two different conductive yarns weaved into narrow fabric in six rows with a pitch of 1.27 mm. For testing purposes, a non-functional printed circuit board was encapsulated with a casting material on the narrow fabric creating rigid–soft (encapsulation–textile) transition areas and accumulated stress for the conductive yarns (Figure 1).

Figure 1. Final sample encapsulation; red markings highlight the transition/high-stress areas.

2. Materials and Methods

The materials are divided into two categories: testing samples and testing machine. The methods explain the testing process in more detail.

2.1. Samples

The testing samples were based on a narrow fabric weaved in conductive yarns. The yarns were integrated into the narrow fabric in 6 rows, with a pitch of 1.27 mm to create a communication bus for the I^2C communication protocol. Two different yarns were tested during this work:

1. Multifilament Ni-plated copper yarn consisting of 6 filaments twisted around a polyester core. The twisted filaments were protected by a Teflon coating. The average resistance of the yarn was 0.8 ohm/m and the outer diameter was 0.60 mm.
2. Multifilament copper-plated steel yarn consisting of 14 twisted filaments. The filaments were protected by a Teflon coating. The average resistance of the yarn was 0.99 ohm/m and the outer diameter was 0.68 mm.

A CO_2 laser was used to create an opening in the yarn for an electrical connection with electronic parts. A rigid PCB (printed circuit board) was glued onto the narrow fabric and connected to the conductive yarns through soldering. The integrated electronics were encapsulated with a polyurethane-based flexible potting compound through the low-pressure injection-molding process. The encapsulation was designed to have a smooth transition from thick to thin. However, the difference in the encapsulation material and the textile flexibility properties created transition points with higher mechanical stress. The final outcome of the sample and the highlighted stress areas are seen in Figure 1.

The conductive-yarn endings on each side were stripped from the Teflon cover. On one side, the conductive-yarn endings were coupled together and soldered to make 3 loops. On the second side, the conductive-yarn endings were soldered on an interposer PCB with female pins (Figure 2).

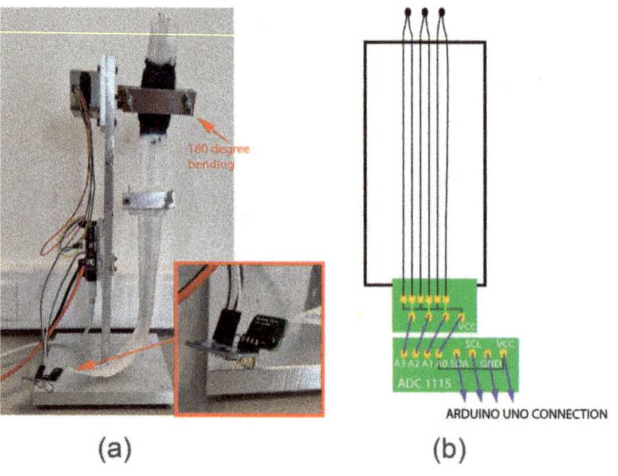

Figure 2. Testing set-up where the motor is at the top left side where also the 180-degree rotation takes place. (**a**) Machine set-up highlighting the connection between ADC and sample and where the 180-degree bending takes place. (**b**) Sample set-up highlighting the connections between interposer board/yarns and interposer board/ADC board.

2.2. Machine

The tester was developed using off-the-shelf components and a specially milled frame (Figure 2). The fixed part of the machine consisted of:

- stepper motor Nema 17;
- Arduino UNO microcontroller together with Motorshield V2 board;
- Adafruit ADC1115 (analogue-to-digital converter) together with male 90-degree pins.

The stepper motor was attached to the top and connected to the Arduino UNO/ Motorshield V2 board. The connections between Motorshield and Arduino board did not allow connecting the sample to the microcontroller. Thus, the ADC board was connected in between to read out the data. The male pins on the ADC board allowed connection and disconnection of the sample easily to the machine.

2.3. Testing Method

A sample with a weight clamp was attached from the (black) encapsulated part to the motor and flexed 180 degrees. Since the yarns were coupled together, one ending of the loop received the power input and the other end was used to read the output. During the test, the samples were flexed 180 degrees for 100,000 cycles or until it was observed that at least two yarn loops' voltage dropped and was not regained for 10 seconds.

The ADC was used to convert digital signals to analogue signals to read every voltage value measured, whether the signal was there or not. If the yarn was damaged in between, the voltage drop would indicate it. Data read-out included: the voltage of the yarns (V), the number of cycles performed, and time (s). The data transmission was read out and saved using PuTTY (an open-source terminal emulator) through the serial port [3]. The data were logged in a CSV file that is easily usable for analysis.

However, resistance rise is often a more desired measurement for failure analysis. Resistance can be measured using Ohm's law by first defining the current (I).

The original resistance of a yarn (R_Y) loop was 1.2 Ω. The voltage input was always 5 V and the original voltage output (V_{out}) was always 4.95 V. Thus, the voltage loss (V_{loss}) at the start was always 50 mV. Based on that, the current was calculated:

$$I = \frac{V_{loss}}{R_Y} = \frac{50\ \text{mV}}{1.2\ \Omega} = 41\ \text{mA} \tag{1}$$

when the resistance of the yarns stays within 10% of the load resistance, the current stays stable and it can be assumed the same. To calculate the load resistance (R_L), the following formula can be used:

$$R_L = R_Y \times \frac{V_{out}}{V_{loss}} = 1.2\ \Omega \times \frac{4.95\ \text{V}}{0.05\ \text{V}} = 120\ \Omega \tag{2}$$

Thus, if the R_Y changes over 12 Ω, then the current will also change. When the yarn resistance changes over the 12 Ω during the tests (voltage drops), the resistance should be calculated accordingly:

$$R_y = \frac{(5\ \text{V} - V_{out}) \times R_L}{V_{out}} \tag{3}$$

Thus, the resistance of the yarn loops can be also plotted during failure analysis (Figure 3b,d).

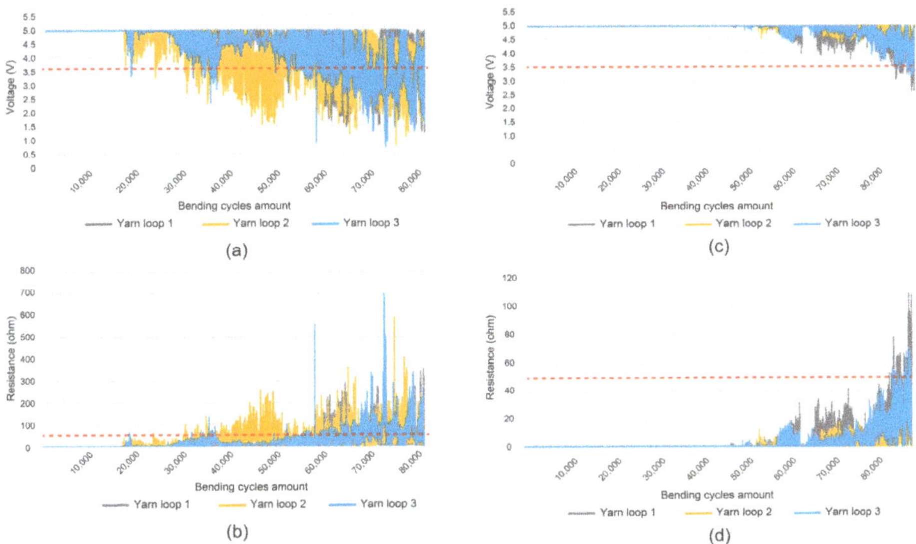

Figure 3. Results with red dotted line highlighting the failure points. (**a**) Yarn 1 (Ni-plated copper yarn) showing voltage drops. (**b**) Yarn 1 (Ni-plated copper yarn) showing resistance changes. (**c**) Yarn 2 (copper-plated steel yarn) showing voltage drops. (**d**) Yarn 2 (copper-plated steel yarn) showing resistance changes.

3. Results

As mentioned earlier, the samples were flexed 180 degrees for 100,000 cycles or until the voltage dropped in at least two yarn loops for 10 seconds after bending. The failure points were based on the final needs of the application. Voltage under 3.5 V or resistance rise to 50 ohms was considered a failure point. The voltage drops are seen in Figure 3a,c comparing two different conductive yarns during the bending tests.

It was also observed that if more than one yarn was breaking during the tests, the second failure appeared quite fast. Moreover, the copper-plated steel yarn (yarn 2) durability was considerably better than the Ni-plated copper yarn (yarn 1). Voltage drops were smaller and more stable, indicating how steel and a larger quantity of filaments provide much better durability than conductive yarns.

Moreover, the data showed how earlier resistance rises were reduced back to original measurements (Figure 3a,c between 20,000 and 30,000 cycles). The stable connections were lost due to the filaments breaking during the bending tests. However, it was still possible for the yarn to regain some connection with other filaments, which resulted in a regain of voltage output as well. Voltage drops increase in size and persistence when filaments break. It was observed that a yarn with a larger number of filaments (yarn 2) regained the original conductivity more, was more stable, and started to break later. Thus, higher numbers of filaments together with less sensitivity to plastic deformation under bending the base material (steel or copper) support stable connections longer.

4. Conclusions and Future Work

This work introduced a bending tester aimed to determine conductive yarns' breaking points. The testing method was also explained by bending two different conductive yarns weaved into a narrow fabric. The results show how the two yarns degraded differently and at what exact point they started to degrade. The tool presents an opportunity to test for a proof-of-concept without ordering expensive machines and tests. Future work will include alternative testing features, such as a slower or faster bending cycle or a different weight clamp.

Author Contributions: Conceptualization, P.V., F.T. and F.B.; methodology, P.V. and F.B.; software, P.V. and F.B.; validation, P.V. and F.B.; formal analysis, P.V., F.B. and J.V.; investigation, P.V. and F.B.; resources, P.V. and F.B.; data curation, P.V.; writing—original draft preparation, P.V.; writing—review and editing, P.V., F.T., F.B. and J.V.; visualization, P.V.; supervision, F.B. and J.V. All authors have read and agreed to the published version of the manuscript.

Funding: This research received no external funding.

Institutional Review Board Statement: Not applicable.

Informed Consent Statement: Not applicable.

Data Availability Statement: Data is contained within the article. Please contact the authors for additional information.

Conflicts of Interest: The authors declare no conflict of interest.

References

1. De Pasquale, G.; Mura, A. Accelerated lifetime tests on e-textiles: Design and fabrication of multifunctional test bench. *J. Ind. Text.* **2018**, *47*, 1925–1943. [CrossRef]
2. Zaman, S.U.; Tao, X.; Cochrane, C.; Koncar, V. Wash Analyses of Flexible and Wearable Printed Circuits for E-Textiles and Their Prediction of Damages. *Electronics* **2021**, *10*, 1362.
3. Bolton, W. I/O Processing. In *Programmable Logic Controllers*, 6th ed.; Bolton, W., Ed.; Newnes: Boston, MA, USA, 2015; Chapter 4; pp. 79–113. [CrossRef]

Proceeding Paper

Ethical Aspects of Health Sensing Applications in E-Textiles †

Sanju Ahuja * and Jyoti Kumar

Department of Design, Indian Institute of Technology Delhi, Hauz Khas, New Delhi 110016, India
* Correspondence: sanju.ahuja@design.iitd.ac.in

† Presented at the 4th International Conference on the Challenges, Opportunities, Innovations and Applications in Electronic Textiles, Nottingham, UK, 8–10 November 2022.

Abstract: The role of e-textiles has been discussed widely for health sensing applications. However, the ethical aspects of health sensing e-textiles have received less attention in the literature. In this contribution, we aim to identify the ethical concerns that emerge from the collection and use of health data from e-textiles. To identify these concerns, we draw upon the literature on health wearables and m-health applications. We propose that four ethical concerns need to be accounted for in the design of e-textiles that collect health data. These are privacy, discrimination, autonomy, and harm. We discuss the need to address these ethical concerns during the design process.

Keywords: health sensing; e-textiles; ethics; surveillance

Citation: Ahuja, S.; Kumar, J. Ethical Aspects of Health Sensing Applications in E-Textiles. *Eng. Proc.* **2023**, *30*, 4. https://doi.org/10.3390/engproc2023030004

Academic Editors: Steve Beeby, Kai Yang, Russel Torah and Theodore Hughes-Riley

Published: 20 January 2023

1. Introduction

E-textiles are being widely developed for health sensing applications, such as smart health, healthy ageing, sports, fitness, and wellness [1]. Sensing technologies can be integrated within e-textiles, such as into the fabric or the yarn, to collect a wide range of health data. E-textiles can communicate with smartphones and computers to collect, display, store, and process physiological information such as heart rate, temperature, breathing, stress, movement, acceleration, or even hormone levels. The use of e-textiles in the health domain can help users monitor health conditions, treat diseases, maximize cognitive and physical function, and promote social engagement and interactions [2]. They can also be used in specialized medical settings to monitor patients' health, postures, and physiological states in real time.

In the health sensing domain, e-textiles have been argued to be the next level of technology after mobile health and wearables [3]. Mobile health (m-health) refers to mobile applications in the health domain, such as fitness apps, weight-watching apps, exercise apps, etc. [4]. Mobile applications collect data through both embedded sensors and self-logging by users. Health wearables refer to devices such as smartwatches and pedometers with sensors for capturing health-related data. The literature has discussed several ethical concerns pertaining to m-health applications and wearables, even though they are voluntarily adopted by millions of users across the globe. However, ethical aspects of e-textiles have received relatively less attention in the literature. This is potentially because e-textiles are an emerging technology that is yet to become as ubiquitous or accessible [4]. However, there is a need to anticipate potential ethical issues in advance. In this paper, we argue that the ethical concerns posed by m-health applications and wearables extend to health sensing applications of e-textiles. To identify these concerns, we draw upon the existing literature from these domains [5,6]. We identify four major ethical concerns, which are privacy, discrimination, autonomy, and harm. We show how these concerns are replicated or exacerbated in the context of e-textiles.

2. Background

Ethical concerns have been discussed widely in the domain of m-health and health wearables. It has been argued that through the use of m-health applications, users become subjects of both surveillance and persuasion [4]. They present their body and their self as open to continual measurement and assessment. Wieczorek et al. [5] conducted a literature review of health wearables and identified areas of opportunities and concerns associated with self-tracking. Sharon [6] discussed the disciplining and disempowering effects of wearables, outlining their impact on values of autonomy, solidarity, and authenticity.

In this paper, we argue that ethical concerns pertaining to m-health and health wearables extend to the health sensing applications of e-textiles as well. Since the data collected by e-textiles are potentially more intimate, accurate, and sensitive, in many cases, these concerns might be exacerbated. However, critical literature has not emerged in this area. Therefore, there is a need to understand the ethical aspects pertaining to the design of e-textiles for health sensing applications.

3. Ethical Aspects of Health Sensing Applications in E-Textiles

In this paper, we draw upon the literature on m-health and health wearables to identify the ethical aspects of health sensing applications in e-textiles. We discuss how these concerns might operationalize in different application contexts. We discuss four ethical aspects of e-textiles: privacy, discrimination, autonomy, and harm (Figure 1).

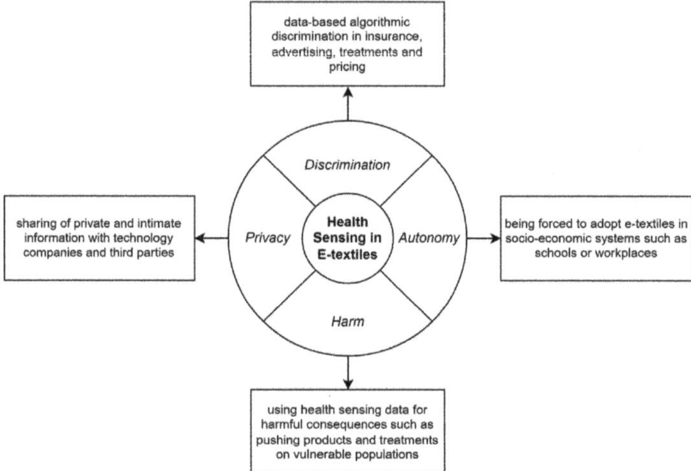

Figure 1. Ethical concerns in health sensing applications in e-textiles.

3.1. Privacy

Inarguably, the most pronounced concern that emerges for health sensing e-textiles is that of privacy [5]. Privacy has emerged as a significant concern across different types of sensing and surveillance technologies. Health sensing technologies can collect a wide range of data, from trivial to intimate. While some health wearables may collect less sensitive data such as the number of steps walked by a user, the data collected by e-textiles are more intimate by design. E-textiles may collect highly sensitive health data such as body temperature, perspiration rates, stress levels and movement patterns [1]. On one hand, these datasets may be better suited for functional purposes. On the other hand, privacy concerns are higher due to the intimate nature of the data. As users adopt e-textiles for several health and fitness purposes, they may inadvertently end up sharing sensitive private information with technology companies, service providers, and third parties.

3.2. Discrimination

The data collected from health-sensing e-textiles poses not only privacy concerns, but also concerns about data-driven discrimination. Health data collected from various sources are often processed for decision-making purposes in different industries. The more sensitive the data are, the more they can be used for inferential decision-making purposes in areas where users least expect it or are not prepared for it. Health data collected by e-textiles can be used for decision-making in markets or in medical setups. When private data are shared with different stakeholders, users may end up being discriminated against by algorithms used in insurance, advertising, treatment, pricing, etc. [7,8]. For example, within insurance, it may so happen that users who are unaware of their own pre-existing conditions are sold insurance plans at the highest prices. Health data may also be used as a justification to discriminate in medical treatments, for example, to prioritize patients in limited-resource settings. In the domain of advertising, algorithms may exploit users' health vulnerabilities to suggest targeted products, which users are tempted to buy out of fear but do not really need.

3.3. *Autonomy*

Another concern that has surfaced prominently in the literature is that of autonomy. The value of autonomy is often invoked to articulate both the benefits and the concerns about health wearables [6]. Health sensing is surrounded by the empowerment narrative, promising greater self-knowledge through numbers [5]. Advertising of health wearables seeks to convince customers to gain control over their health, weight, and sleep, facilitated by troves of data. However, the empowerment narrative is accompanied by perceptions of individual responsibility. In the workplace, employers are encouraging employees to adopt wearables [9]. Insurance companies commonly offer discounts to customers who self-track [8]. In such arrangements, the line between voluntary and compulsory participation often blurs [7]. The normalization of e-textiles for health sensing and their integration within socio-economic systems may force users into involuntary participatory arrangements. Users may be forced to adopt e-textiles in schools and workplaces, akin to the current push towards wearables [9]. Another autonomy-related critique of health sensing concerns the value of authenticity. Data-driven approaches to health have been criticized in the literature as unidirectional and reductive [5,6]. It has been argued that these approaches prevent individuals from exploring alternate means of health management.

3.4. *Harm*

Lastly, the data collected through e-textiles may be used intentionally or unintentionally to cause harm. For example, e-textiles with sensors that detect users' stress levels may share their data with advertising companies, promoting the sale of addictive medication while a user is in their most vulnerable state. Health sensing data may be used as a justification to deny users medical treatments. E-textiles may also cause unintended harm. They may inadvertently expose users to information about underlying medical conditions without them seeking out this information. They may cause users undue mental distress with false positives about underlying medical conditions. In contrast, they may fail to detect an ongoing medical emergency, such as a heart attack, giving users a false sense of comfort even when their symptoms tell them otherwise.

4. Discussion and Conclusions

The ethical aspects highlighted in this paper apply to nearly all use cases of health-sensing e-textiles. There is a need for designers to actively mitigate these concerns. For this purpose, it is important that designers are able to anticipate these concerns in advance, during the design activity, a process termed moral imagination [10]. Moral imagination allows designers to anticipate ethical concerns using ethics frameworks such as the one presented in this paper. Designers can systematically assess whether their sensing applications violate users' privacy or autonomy, or whether they can be used to perpetuate discrimination or cause any harm. Addressal of these concerns is important to build users' trust in technology, especially when several technology products are facing a 'crisis of ethics' [11]. We believe that this contribution is a timely one to initiate a discussion on the ethical aspects of e-textiles, and foreground ethical concerns in the design activity.

Author Contributions: This research was conducted as part of the doctoral dissertation of S.A. and was supervised by J.K. All authors have read and agreed to the published version of the manuscript.

Funding: This research was supported by the Prime Minister Doctoral Research Fellowship granted by the Ministry of Human Resource Development, Government of India.

Institutional Review Board Statement: Not applicable.

Informed Consent Statement: Not applicable.

Data Availability Statement: Not applicable.

Conflicts of Interest: The authors declare no conflict of interest.

References

1. Fernández-Caramés, T.M.; Fraga-Lamas, P. Towards the Internet of Smart Clothing: A Review on IoT Wearables and Garments for Creating Intelligent Connected E-Textiles. *Electronics* **2018**, *7*, 405. [CrossRef]
2. Yang, K.; Isaia, B.; Brown, L.J.E.; Beeby, S. E-Textiles for Healthy Ageing. *Sensors* **2019**, *19*, 4463. [CrossRef] [PubMed]
3. Mertz, L. E-Textiles for Health Monitoring: Off to a Slow Start, but Coming Soon. *IEEE Pulse* **2020**, *11*, 11–20. [CrossRef] [PubMed]
4. Lupton, D. M-health and health promotion: The digital cyborg and surveillance society. *Soc. Theory Health* **2012**, *10*, 229–244. [CrossRef]
5. Wieczorek, M.; O'Brolchain, F.; Saghai, Y.; Gordijn, B. The ethics of self-tracking. A comprehensive review of the literature. *Ethics Behav.* **2022**. [CrossRef]
6. Sharon, T. Self-Tracking for Health and the Quantified Self: Re-Articulating Autonomy, Solidarity, and Authenticity in an Age of Personalized Healthcare. *Philos. Technol.* **2017**, *30*, 93–121. [CrossRef]
7. Ahuja, S.; Kumar, J. Surveillance based Persuasion: The Good, the Bad and the Ugly. In Proceedings of the 4th International Conference on Computer-Human Interaction Research and Applications (CHIRA 2020), Budapest, Hungary, 5–6 November 2020; pp. 120–127. [CrossRef]
8. McFall, L. Personalizing solidarity? The role of self-tracking in health insurance pricing. *Econ. Soc.* **2019**, *48*, 52–76. [CrossRef] [PubMed]
9. Chung, C.; Gorm, N.; Shklovski, I.A.; Munson, S. Finding the Right Fit: Understanding Health Tracking in Workplace Wellness Programs. In Proceedings of the 2017 CHI Conference on Human Factors in Computing Systems (CHI 2017), Denver, CO, USA, 6–11 May 2017; pp. 4875–4886. [CrossRef]
10. Ahuja, S.; Kumar, J. Assistant or Master: Envisioning the User Autonomy Implications of Virtual Assistants. In Proceedings of the 4th Conference on Conversational User Interfaces (CUI 2022), Glasgow, UK, 26–28 July 2022. [CrossRef]
11. Wessel, M.; Helmer, N. A Crisis of Ethics in Technology Innovation. *MIT Sloan Management Review*. 2020. Available online: https://sloanreview.mit.edu/article/a-crisis-of-ethics-in-technology-innovation/ (accessed on 2 December 2022).

Proceeding Paper

Simultaneous Breathing and ECG Measurements with e-Knits †

Kristel Fobelets *, Christoforos Panteli and Ghena Hammour

Department of Electrical and Electronic Engineering, Imperial College London, Exhibition Road, London SW7 2BT, UK
* Correspondence: k.fobelets@imperial.ac.uk
† Presented at the 4th International Conference on the Challenges, Opportunities, Innovations and Applications in Electronic Textiles, Nottingham, UK, 8–10 November 2022.

Abstract: Simultaneous recording of breathing and heart rate signals was carried out on a healthy volunteer with a fully knitted, non-sports-type garment. Breathing was recorded using two knitted respiratory inductive plethysmography (RIP) sensors. Electrocardiogram (ECG) recordings were obtained from three knitted electrodes. The knitted garment design was customised for the specific requirements of RIP and ECG by adapting the needle size and/or introducing knit-in-elastic in the sensor areas. RIP was read out using an in-house-developed cross-coupled complementary oscillator circuit. The ECG was recorded using the commercial OpenBCI board. The sensors produced excellent signal quality that allowed for simple signal processing to extract information on heart and breathing rates, showing good correlation between the two.

Keywords: respiratory inductive plethysmography; ECG; e-wearable sensor; knitting; oscillator

1. Introduction

Among service prices, population growth and aging, chronic conditions play a major role in substantially increasing healthcare costs. Wearable technology is attractive for continuous health monitoring and thus reduces the costs associated with doctor–patient interactions. Knitting offers the possibility to integrate different sensors for non-invasive and real-time breathing and heart rate monitoring in wearable garments. This can be achieved by tracking changes in chest/abdomen circumference using transduction methods such as resistance changes [1,2] or inductance changes of a coil wound around the body [3]. In most cases, RIP (respiratory inductive plethysmography) is implemented in elastic belts that are strapped around the chest and/or the abdomen, shown in Figure 1a. For ECG, gelled electrodes are normally taped on the body. A more wearable implementation comes in the form of dry knitted electrodes [4]. In our implementation, knitted RIP and ECG were optimised in conjunction with the garment to increase sensitivity and reduce motion artifacts, illustrated in Figure 1b.

Citation: Fobelets, K.; Panteli, C.; Hammour, G. Simultaneous Breathing and ECG Measurements with e-Knits. *Eng. Proc.* **2023**, *30*, 5. https://doi.org/10.3390/engproc2023030005

Academic Editors: Steve Beeby, Kai Yang, Russel Torah and Theodore Hughes-Riley

Published: 20 January 2023

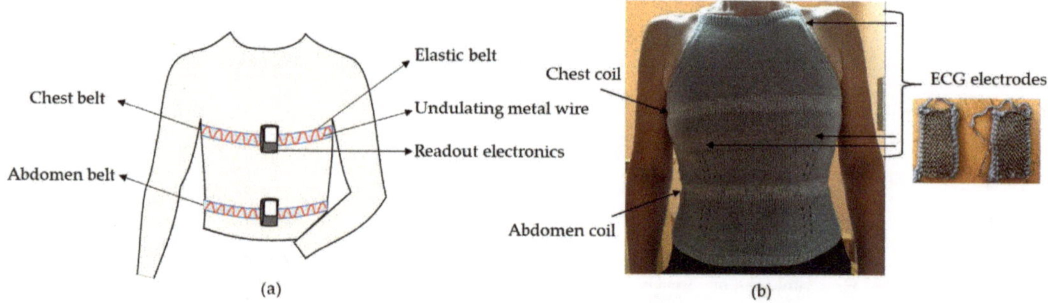

Figure 1. (**a**) Classical RIP implementation using elastic belts with one metal winding strapped around the chest and abdomen. (**b**) Knitted implementation of RIP in a halter top, on the chest and on the abdomen. The ECG electrodes and their positions on the inside of the garment are shown with arrows. Inset show the knitted electrodes.

2. Materials and Methods

2.1. Knitted Breathing Sensors

In our previous work, RIP sensors were integrated into a garment by a thin, circular knitted, insulated metal wire with yarn [5], achieving ultra-wearability and higher sensitivity proportional to the number of knitted rows with metal [6]. Figure 1b shows the implementation of two knitted coils, at the chest and the abdomen level. The RIP sensors were knitted using a needle size appropriate for the yarn, allowing the knit's natural elasticity to accommodate the stretch when inhaling. Knit-in-elastic is added to force the knit to return to its minimum circumference when exhaling.

2.2. RIP Read-Out Electronics

A complementary cross-coupled pair oscillator translates the coil's inductance to frequency [7]. For a wide range of coil dimensions to be recorded, the oscillator is followed by a rail-to-rail comparator that converts the sine-wave oscillations to a rectangular waveform. An esp32 microcontroller counts the frequency and logs the data to a micro-SD card [8].

2.3. Knitted ECG Electrodes

The ECG electrodes were knitted using Ag-coated polyester thread with a size equivalent to commercial pads. The different implementations are given in Table 1.

Table 1. Five different types of knitted electrodes. The four on the right were knitted by hand.

Electrode	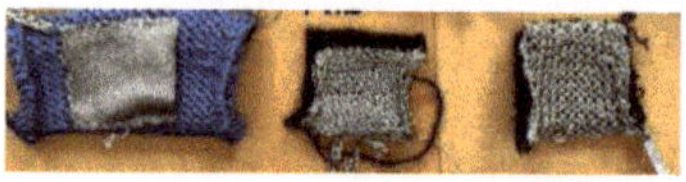				
Needle size (mm)	*	1	1.75	2	2.5
Number of threads +	*	1	2	3	4

* Commercially knitted Shieldex fabric [9]. + Shieldex threads (235/36 × 2 HCB in catalogue [9]).

By knitting different yarn thicknesses and adapting the needles' size, the roughness of the rib side of the electrodes can be controlled while maintaining a similar stitch density, unlike in the previous work [10]. The rib side of the knit is placed against the skin to reduce movement artifacts and improve signal quality. To decrease the movement artifacts in the ECG signals further, the regions in the knitted garment where the electrodes are sewn in, are knitted with a smaller needle size. This reduces the elasticity of the knit in those areas, reducing movement against the body. The ECG signals were recorded using the OpenBCI board [11]. The quality of the electrodes was defined by comparing the mean and median signal and signal frequency histogram that must be non-Gaussian. The electrode with 3 Shieldex threads and 2 mm needles gave the best performance.

3. Results

3.1. Breathing of the Volunteer

The performance parameters for a range of coils are given in Table 2. A larger number of windings N results in larger sensitivity s and smaller current i_{coil}. Thus, although Q is low for high N, a reduced i_{coil} is better for user safety. Using coils with $N = 8$ in the knit of Figure 1, different breathing patterns, normal, slow and fast, were recorded from a healthy volunteer and are given in Figure 2.

Table 2. Characteristics of the knitted coil: c circumference, N number of windings, s sensitivity, f_{osc} oscillator frequency, Q quality factor and i_{coil} the RMS current through the coil at f_{osc}.

Needle Size (mm)	c (cm)	N	s (μH/cm)	f_{osc} (MHz)	Q	i_{coil} (mA)
1.75	27.5	13	1.20	3.3	72	1.2
~1.75 *	78.5	60	13.20	0.2	15	0.5
2	90.5	8	0.56	2.1	26	0.9

* Machine knitted on a Kniterate [12] machine by Ecoknitware [13].

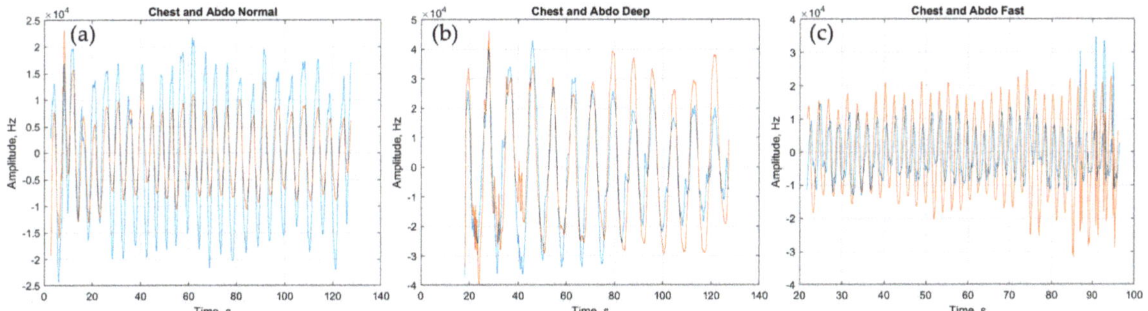

Figure 2. Frequency variations of the chest (blue) and abdomen (orange) (colour available online) coils due to breathing. (**a**) Normal breathing, (**b**) deep and slow breathing and (**c**) fast breathing.

Breathing parameters (Table 3) were extracted using signal processing in MATLAB. The signal amplitude is related to the effort from the chest and abdomen. The phase difference during fast breathing mirrors minor hyperventilation and is the result of the increased work of breathing and the use of accessory muscles.

Table 3. Breathing parameters. BPM: breaths per minute, T_i/T_{tot} the ratio of the inhalation time to the total time of one breath, $|A|$ the mean amplitude of one breath and $\Delta\phi$ the phase difference between the signal from the chest and the abdomen.

	Chest Coil			Abdomen Coil			Phase
Breathing Type	**BPM**	T_i/T_{tot}	$\|A\|$ **(kHz)**	**BPM**	T_i/T_{tot}	$\|A\|$ **(kHz)**	$\Delta\phi$ **(°)**
Normal	12	0.44	17	13	0.41	34	3
Deep	7	0.56	61	7	0.43	52	4
Fast	29	0.47	35	29	0.46	20	49

3.2. ECG on the Volunteer

Figure 3 shows 10 s ECG snapshots of 1 min measurements using the dry knitted contacts as implemented in Figure 1b during normal, fast and slow breathing. Typical signal parameters were extracted in MATLAB, as given in Table 4. In Figure 3, we observe that the amplitude of the R-peaks (the peaks with the largest amplitude) is modulated by the breathing signal. Breathing in, increases the resistance of the chest and breathing out decreases it. The breathing rate associated to this modulation is similar to that reported in Table 3.

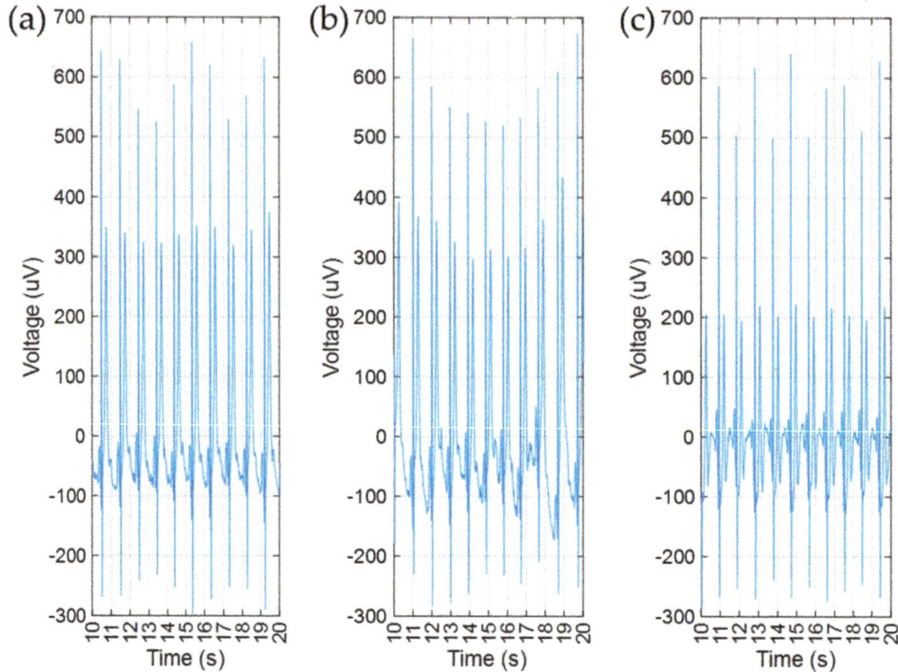

Figure 3. 10 s ECG recordings for (**a**) normal, (**b**) deep and (**c**) fast breathing. All recorded using dry knitted electrodes.

Table 4. ECG parameters. *R*: distance between R-peaks, σ: standard variation on the R-peak position and HR: heart rate in beats per second (bps).

	Horizontal Electrodes			Vertical Electrodes			
Breathing Type	***R* (s)**	**σ (s)**	**RH (bps)**	***R* (s)**	**σ (s)**	**RH (bps)**	**BPM**
Normal	0.98	0.225	61	0.98	0.225	61	12
Deep	1.02	0.210	59	1.02	0.240	59	6
Fast	0.94	0.156	64	1.00	0.234	60	20

4. Conclusions

Knitted RIP sensors and ECG electrodes were implemented in a garment and were used to record breathing and ECG signals simultaneously. This implementation gave good quality recordings and health parameters in a relatively relaxed fitting garment when the wearer sat still.

Author Contributions: Conceptualization, K.F.; methodology, K.F., C.P. and G.H.; software, K.F. and G.H.; validation, K.F.; formal analysis, K.F.; investigation, K.F. and C.P.; resources, K.F.; data curation, K.F. and C.P.; writing—original draft preparation, K.F.; writing—review and editing, C.P.; visualization, K.F.; supervision, K.F.; project administration, K.F.; funding acquisition, K.F. All authors have read and agreed to the published version of the manuscript.

Funding: This research received no external funding.

Institutional Review Board Statement: Ethical review and approval were waived for this study as the dataset is not related to clinical trials.

Informed Consent Statement: Informed consent was obtained from all subjects involved in the study.

Data Availability Statement: Data can be requested from K.F.

Conflicts of Interest: The authors declare no conflict of interest.

References

1. Liang, A.; Stewart, R.; Bryan-Kinns, N. Analysis of Sensitivity, Linearity, Hysteresis, Responsiveness, and Fatigue of Textile Knit Stretch Sensors. *Sensors* **2019**, *19*, 3618. [CrossRef] [PubMed]
2. Wang, Y.; Ali, S.; Wijekoon, J.; Hugh Gong, R.; Fernando, A. A wearable piezo-resistive sensor for capturing cardiorespiratory signals. *Sens. Actuators A Phys.* **2018**, *282*, 215–229. [CrossRef]
3. Ngo, H.T.; Nguyen, C.V.; Nguyen, T.M.H.; Van Vo, T. A Portable Respiratory Monitor Using Respiratory Inductive Plethysmography. In Proceedings of the 4th International Conference on Biomedical Engineering in Vietnam, Ho Chi Minh City, Vietnam, 8–12 January 2012; IFMBE Proceedings; Springer: Berlin/Heidelberg, Germany, 2013; Volume 49. [CrossRef]
4. Le, K.; Narayana, H.; Servati, A.; Bahi, A.; Soltanian, S.; Servati, P.; Ko, F. Electronic textiles for electrocardiogram monitoring: A review on the structure–property and performance evaluation from fiber to fabric. *Text. Res. J.* **2022**, 1–33. [CrossRef]
5. Fobelets, K. Knitted coils as breathing sensors. *Sens. Actuators A Phys.* **2020**, *306*, 111945. [CrossRef]
6. Fobelets, K.; Panteli, C.; Hammour, G. E-knits for bio-signal recordings. In Proceedings of the 4th International e-Textiles Conference, Nottingham, UK, 9–10 November 2022.
7. Kiener, K.; Anand, A.; Fobelets, W.; Fobelets, K. Low Power Respiration Monitoring Using Wearable 3D Knitted Helical Coils. *IEEE Sens. J.* **2022**, *22*, 1374–1381. [CrossRef]
8. Panteli, C.; Fobelets, K. Complementary cross-coupled LC oscillator for Respiratory Inductive Plethysmography readout. In Proceedings of the 4th International e-Textiles Conference, Nottingham, UK, 9–10 November 2022.
9. Shieldex®. Available online: https://www.shieldex.de/en/products_categories/fibers-yarns/ (accessed on 18 January 2023).
10. Euler, L.; Guo, L.; Persson, N.-K. Textile Electrodes: Influence of Knitting Construction and Pressure on the Contact Impedance. *Sensors* **2021**, *21*, 1578. [CrossRef] [PubMed]
11. Low-Cost Biosensing Starter Kit | OpenBCI. Available online: https://openbci.com/ (accessed on 30 November 2022).
12. Kniterate | The Digital Knitting Machine. Available online: https://www.kniterate.com/ (accessed on 30 November 2022).
13. EcoKnitware. Available online: https://ecoknitware.com/ (accessed on 30 November 2022).

Proceeding Paper

Hit the Ground Running—Wearable Sensors to Measure Foot Plantar Pressure †

Niamh Saunders [1], Karnasooriya Ragalage Sanjaya Dinuwan Gunawardhana [2,3,*], Luz Alejandra Magre Colorado [2], Sonal Santosh Baberwal [2] and Shirley Coyle [2,3,*]

1 School of Mechatronic Engineering, Dublin City University, Glasnevin, D09Y074 Dublin, Ireland
2 School of Electronic Engineering, Dublin City University, Glasnevin, D09Y074 Dublin, Ireland
3 Insight Centre for Data Analytics, Dublin City University, Glasnevin, D09Y074 Dublin, Ireland
* Correspondence: sanjaya.gunawardhana2@mail.dcu.ie (K.R.S.D.G.); shirley.coyle@dcu.ie (S.C.)
† Presented at the 4th International Conference on the Challenges, Opportunities, Innovations and Applications in Electronic Textiles, Nottingham, UK, 8–10 November 2022.

Abstract: Flexible pressure sensors can be used to predict possible injuries and inform the wearer about their foot posture and landing positions during both walking and running. This work focuses on designing and producing capacitive pressure sensors and integrating them into a smart insole. The suitability of sensors was assessed using a mechanical test rig to measure the change of capacitance under different loads. The effects of introducing micropores into the dielectric layer using two fabrication methods were analysed. The results obtained imply that the use of micropores has the potential to increase the sensitivity and improve the response time of the sensors.

Keywords: capacitive pressure sensor; wearable electronics; foot plantar pressure; smart insole

Citation: Saunders, N.; Gunawardhana, K.R.S.D.; Magre Colorado, L.A.; Baberwal, S.S.; Coyle, S. Hit the Ground Running—Wearable Sensors to Measure Foot Plantar Pressure. *Eng. Proc.* **2023**, *30*, 6. https://doi.org/10.3390/engproc2023030006

Academic Editors: Steve Beeby, Kai Yang, Russel Torah and Theodore Hughes-Riley

Published: 20 January 2023

1. Introduction

Foot plantar pressure is an essential parameter for sports and healthcare applications. Pressure sensors, integrated into force plates and shoe insoles, are amongst the main tools used to measure foot plantar pressure which can analyse various movements performed by the user. Insole pressure sensors have a broad spectrum of multiple applications, for instance, in gait and motion analysis, rehabilitation, sports training, step counting, and detection of loss of balance [1]. These applications need to measure pressure in different parts of the sole in order to identify foot posture related to the wearer's movement activities.

This creates a need for robust pressure sensors that can be integrated using textile manufacturing techniques and ensure that the device has suitable properties to be worn inside the shoe. As with developing any wearable sensor, there are some properties to consider. Stretchability is one of the essential characteristics of a wearable sensor due to the naturally irregular surface of the skin, and changes can be presented as a consequence of the normal movement [2].

This work demonstrates an initial development of a low-cost pressure sensor that can be embedded into an insole to continuously measure foot plantar pressure over time, giving the user a more natural movement and therefore obtaining better readings. This, in turn, will lead to better measures for injury prevention. The developed sensor employs a conductive fabrics and silicone elastomer to fulfil the stretchability property.

2. Materials and Method

In this study, parallel plate capacitive pressure sensors were created using a layered structure of conductive fabrics with a middle dielectric elastomer layer. Two types of conductive fabrics were used as conductive layers, a woven (70% polyester, 16% copper and 14% nickel $-0.2\ \Omega$ per 20 cm) layer and a knitted fabric (83% nylon and 17% silver $-1.4\ \Omega$ per 20 cm) layer. Ecoflex 00-30 was used as the middle dielectric layer. In addition,

Chitosan (medium molecular weight) and Acetic acid (99.7%) were purchased from Sigma-Aldrich to prepare the insulation layer.

Three aluminium moulds were created to fabricate the dielectric layer of the sensors, with dimensions of 0.041 m × 0.041 m × 800 μm. Two moulds were made with vertical columns (diameter 600 μm each, with 5 and 9 columns) to insert vertical pores into the dielectric layer (Figure 1a), while the remaining one was without columns.

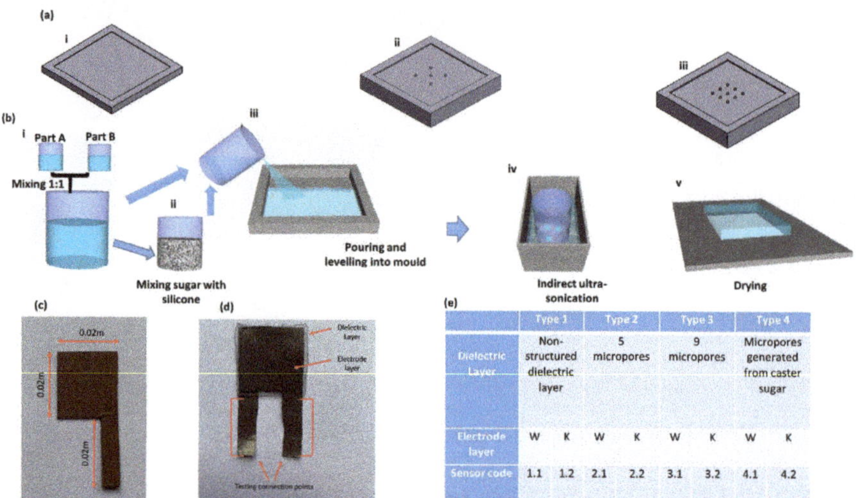

Figure 1. Fabrication process and surface examination of sensors. (**a**) Moulds prepared for dielectric layer fabrication, showing moulds (i) without vertical columns, (ii) with 6 vertical columns and (iii) with 9 vertical columns; (**b**) dielectric layer fabrication, showing (i) mixing Ecoflex part A and Part B 1:1, (ii) mixing with sugar granules, (iii) pouring into the prepared mould, (iv) indirect ultra-sonication and (v) final dielectric layer; (**c**) electrode dimensions, template for fabrication; (**d**) fabricated capacitive pressure sensor with conductive fabric electrode layers and a middle dielectric elastomer layer; (**e**) fabricated sample categorization.

Ecoflex 00-30 silicone solutions A and B were mixed in a 1:1 ratio. The mixture was poured into moulds and kept inside the oven at 70 °C for two hours to cure the silicone (Figure 1b). Cured silicone was extracted carefully for the secondary process. A second approach to integrating porosity within the dielectric was taken, using microcrystals that could be dissolved after curing. The process was repeated to fabricate the dielectric layer by adding caster sugar in the initial mixing phase. A 20 g volume of the Ecoflex 00-30 solution was mixed with 5 g of caster sugar. Once the solution was homogenous, it was poured into the mould with no holes and kept in a 70 °C oven for 2 h. Indirect ultra-sonication dissolved the sugar granules and created a microporous structure. Developed dielectric layers were cut into 21 mm × 21 mm dimensions. Eight samples were developed using knitted and woven fabrics separately, along with four dielectric layers (Table 1, Figure 1e). After that, 1.5 g of chitosan was mixed with 2% Wt % Acetic solution 50 mL and stirred using magnetic stirring for 4 h at 40 °C. The solution was poured into a casting mould and kept for 6 h to dry at 40 °C. An insulation layer made with chitosan was made as a pouch, and the fabricated sensor was put into the pouch before characterizing.

Table 1. Slope and R^2 values calculated for all sensors.

Sensor Type	Slope	R^2 Value
Type 1.1	4.21	0.95
Type 1.2	3.39	0.95
Type 2.1	3.44	0.82
Type 2.2	4.92	0.98
Type 3.1	3.17	0.97
Type 3.2	3.88	0.98
Type 4.1	3.05	0.98
Type 4.2	2.05	0.96

3. Results

SEM imaging was carried out with Jeol JSM-IT 100 InTouchScope SEM at the top surface and tilting angle of 70°. From the SEM image, the thickness of the dielectric layer was measured as 780.729 µm. Additionally, the average size of the sugar granule was 616.519 × 203.338 µm (Figure 2a,b). SEM images confirmed that the ultrasonic process caused all the sugar granules in the dielectric layers to be dissolved, creating a pore with 521.265 × 255.178 µm. A Keysight U1701B handheld capacitance meter and Univert CellScale Machine were used to test the fabricated capacitive pressure sensors (Supplementary note S1). Using the CellScale machine, the force was applied by changing the displacement in true strain function with stretch, recovery and holding for 10 s, respectively. Stretch magnitude was changed between 2%, 3% and 4% displacement. Each test comprises five cycles (supplementary note S2). Graphs were plotted between constant displacement and average capacitance for all sensors (Figure 2c). Line of best fit and regression analysis was carried out to find the relationship between the displacement and capacitance. The recorded slope and coefficient of determination (R^2 value) values for each sensor are given in Table 1.

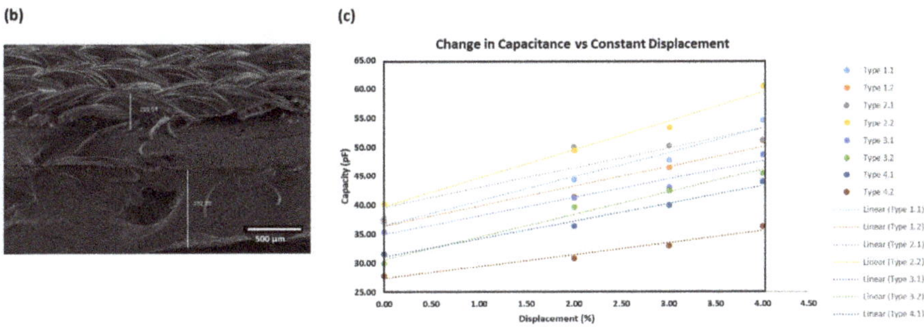

Figure 2. (**a**) SEM images of sugar granular mixed dielectric sample (i) before and (ii) after washing, (**b**) SEM of fabricated sensor (**c**) Comparison of 8 sensors capacity change with respect to displacement.

4. Discussion

The slope of the graph indicates the sensitivity of each sensor, while the R^2 value indicates the linear relationship between the displacement and the capacity. Sensor type 1 is the non-structured dielectric layer. This sensor acted as the control throughout the project to compare the different dielectric layers.

Sensor type 2 and type 3 are made with vertical pore structures. The experimental results indicate that the sensor type 2 woven electrode has a low sensitivity and linearity, and that the knit electrode has a high standard deviation for all three constant displacements individually. On the other hand, sensor type 3 with the woven electrode had an average performance, while the knit electrode demonstrated a high sensitivity and linearity along with a high standard deviation (supplementary note S3). Further investigations are required to prove the effect of adding more vertical pore structures over the surface.

Sensor type 4 has a microporous dielectric layer created from sugar granules. This sensor has a low standard deviation and a moderate slope. This shows that the sensor is moderately sensitive to a change in capacitance when the constant displacement percentage increases.

$$C = \varepsilon_0 \varepsilon_r \frac{A}{d} \quad (1)$$

The formula (Equation (1)) for a parallel plate capacitor's capacitance (C) can be broken into four variables (ε_r, d, A: relative permittivity, thickness and area of the dielectric material, respectively and ε_0: permittivity of the vacuum). Air has an ε_r of 1.0005 while Ecoflex 00-30 has ε_r of 2.8 [3]. This indicates that an increase in air gaps can reduce ε_0 ε_r, thus reducing the capacitance. The reduction of the capacitance of sensor type 4 gives evidence of this behaviour. This could indicate that a further increase in micropores in a smaller dielectric area could reduce the capacitance. However, when pressure is applied to the sensor, the air gap between the two layers is reduced. This causes a more significant deformation, resulting in a higher sensitivity in the sensor [4]. Based on the limitations of the measurement techniques observed in this study, the further development of the testing conditions and use of the more sensitive LCR meter and high frequency, high force linear actuation system would be needed to allow for accurate comparisons to be between the different approaches to fabricated capacitive pressure sensors.

5. Conclusions

This work presents the development of a capacitance-based flexible pressure sensor using textile-based materials. Two methods of adapting the properties of the dielectric layer were tested by increasing the porosity of this layer. The porosity of the layer was confirmed using SEM imaging. The results obtained imply that sensitivity and response time can be improved by implementing microporous structures in the dielectric layer. The study used Chitosan as an insulation layer around the pressure sensor to shield from external interference with the sensor. The use of chitosan enhances the wearable requirements of a sensor placed in contact with the body in terms of biocompatibility and antibacterial properties. The developed sensor has the potential to be integrated into a smart insole which would contribute towards better indicators for injury prevention, rehabilitation progress, fitness assessment and sports performance.

Supplementary Materials: The following supporting information can be downloaded at: https://www.mdpi.com/article/10.3390/engproc2023030006/s1, (Supplementary note S1–S3).

Author Contributions: Conceptualization, Methodology and Analysis, N.S., K.R.S.D.G. and S.C.; writing—original draft preparation, K.R.S.D.G., L.A.M.C., S.S.B.; writing—review and editing, supervision, project administration and Funding acquisition S.C. All authors have read and agreed to the published version of the manuscript.

Funding: This research is funded by Insight Centre for Data Analysis at Dublin City University (SFI/12/RC/2289_P2).

Institutional Review Board Statement: Not applicable.

Informed Consent Statement: Not applicable.

Data Availability Statement: Results are available in this manuscript and additional data can be provided on request.

Acknowledgments: Authors acknowledge Eoin Tuohy, Liam Lawlor and Michael May for the technical support and guidance given through the device fabrication and characterization. The SEM was carried out at the Nano Research Facility in Dublin City University which was funded under the Programme for Research in Third Level Institutions (PRTLI) Cycle 5. The PRTLI is co-funded through the European Regional Development Fund (ERDF), part of the European Union Structural Funds Programme 2011–2015.

Conflicts of Interest: The authors declare no conflict of interest.

References

1. Ngueleu, A.M.; Blanchette, A.K.; Bouyer, L.; Maltais, D.; McFadyen, B.J.; Moffet, H.; Batcho, C.S. Design and Accuracy of an Instrumented Insole Using Pressure Sensors for Step Count. *Sensors* **2019**, *19*, 984. [CrossRef]
2. Liu, Y.; Bao, R.; Tao, J.; Li, J.; Dong, M.; Pan, C. Recent Progress in Tactile Sensors and Their Applications in Intelligent Systems. *Sci. Bull.* **2020**, *65*, 70–88. [CrossRef] [PubMed]
3. Cholleti, E.R.; Stringer, J.; Assadian, M.; Battmann, V.; Bowen, C.; Aw, K. Highly Stretchable Capacitive Sensor with Printed Carbon Black Electrodes on Barium Titanate Elastomer Composite. *Sensors* **2019**, *19*, 42. [CrossRef]
4. Chen, W.; Yan, X. Progress in Achieving High-Performance Piezoresistive and Capacitive Flexible Pressure Sensors: A Review. *J. Mater. Sci. Technol.* **2020**, *43*, 175–188. [CrossRef]

engineering proceedings

MDPI

Proceeding Paper

Automotive Seat Occupancy Sensor Based on e-Textile Technology †

Marc Martínez-Estrada *, Ignacio Gil and Raúl Fernández-García

Department of Electronical Engineering, Universitat Politecnica de Catalunya, 08222 Terrassa, Spain
* Correspondence: marc.martinez.estrada@upc.edu

† Presented at the 4th International Conference on the Challenges, Opportunities, Innovations and Applications in Electronic Textiles, Nottingham, UK, 8–10 November 2022.

Abstract: A woven capacitive sensor is presented as a novel automotive seat occupancy sensor. The woven capacitive sensor is developed as a response to one of the most difficult challenges that the automotive sector has faced the past few years: seat occupancy sensor false positives. The designed sensor demonstrates the error-free sitting detection of a person by detecting the change of the permittivity around it, providing a distinction between a person and an object with the same weight.

Keywords: woven; sensor; occupancy; automotive; textile

1. Introduction

Automotive seat occupancy sensors have been used for many years by car companies to detect the presence of a seated person and activate the seatbelt sound alarm. Force-sensitive resistors (FSR) [1] have been the most common sensors used by automotive companies to detect the presence of a person on the car seats. FSR are piezoresistive sensors that detect a change in the pressure on the seat surface provoking a modification in the resistance. A person is detected by setting a minimum threshold pressure that is set up on the sensors to avoid the seatbelt sound alarm from being activated by a light object. However, there are still false positives provoked by heavy objects that exceed that threshold.

In [2], a flexible resistive sensor it is presented that is produced by a layer disposition of different materials, as an alternative option to FSR. The flexible sensor presented could be integrated into the seat fabric or just underneath it. However, as the working principle of the sensor is resistance variation by the change in the applied force, the possibility to have false positives caused by heavy objects still exists.

There have been attempts to change the nature of the FSR sensor for the capacitive sensor. The aim is to avoid false detection provoked by heavy objects. In [3], a capacitive seat sensor for multiple occupancy detection is presented. The capacitive sensor presented varies its value depending on the person characteristics when it is sat on. The sensor consists of an aluminium foil integrated on the seat. Two capacitances are measured: one of them is measured between bottom part of the car chassis and the aluminium foil, and the other one is measured between the top part of the chassis and the aluminium foil. Both measurements take into account wide spaces, which could imply more possibilities to produce false detections.

In this paper, the automotive sensor presented is a step forward. The capacitive sensor is integrated into the seat fabric. As both of the sensor electrodes are in the textile, the capacitance is affected by closer environment and contact, avoiding the measurement of long areas which could imply new false positives.

The paper is organized as follows: Section 2 covers the materials, methods and characterisation details. In Section 3, the experimental results are presented and discussed. Finally, in Section 4, the conclusions are summarized.

Citation: Martínez-Estrada, M.; Gil, I.; Fernández-García, R. Automotive Seat Occupancy Sensor Based on e-Textile Technology. *Eng. Proc.* **2023**, *30*, 7. https://doi.org/10.3390/engproc2023030007

Academic Editors: Steve Beeby, Kai Yang, Russel Torah and Theodore Hughes-Riley

Published: 20 January 2023

2. Materials and Methods

The proposed automotive seat occupancy sensor is based on an interdigital structure which is integrated into the woven fabric structure. Figure 1 shows the integrated sensor structure. The textile capacitive sensor was built using a commercial Bekaert conductive yarn. The yarn was produced by the ring yarn method, where the conductive stainless steel fibbers were mixed with polyester fibbers in a proportion of 40/60%. The resultant yarn has a linear resistance of 50 Ω/cm.

Figure 1. Interdigital structure introduced into the fabric structure.

To manufacture the textile capacitive sensor, the interdigital structure was built during the weave process of the fabric using a Dornier LWV8/J 71 weaving machine moved by a Jacquard Stäubli LX1600B. Some warp yarns were substituted by conductive yarns to prepare the vertical electrodes. The weft yarns were introduced normally by the weft system. The woven fabric was formed by cotton yarns in the warp and cotton–polyester mix yarns in the weft.

The textile capacitive sensor is incorporated over a car seat, as it can be seen in Figure 2. One of the sensor sides has two snap connections where the sensor is connected to the microcontroller, which measures the sensor value. The microcontroller performs a common charge–discharge method to evaluate the capacitance value of the sensor; this method has been further explained and studied in previous works [4].

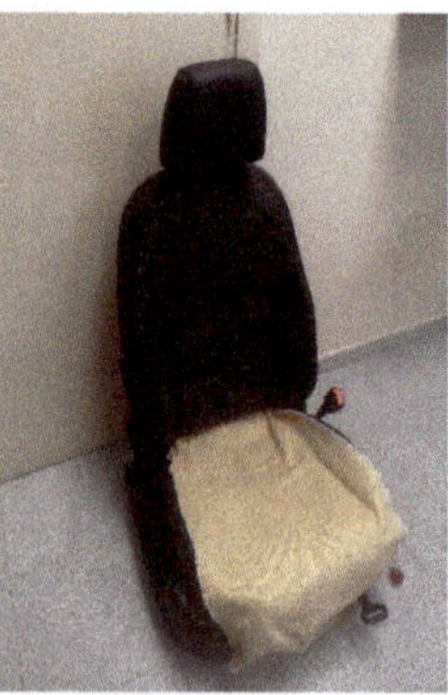

Figure 2. Textile capacitive sensor incorporated over a car seat.

To study the behaviour of the textile capacitive sensor over the car seat, three cases have been evaluated: a person, a heavy object and a bottle of water. The general procedure is similar for every case. The test starts with the seat unoccupied. Then, after a stabilization time, the studied person or object is placed over the sensor for 10 s. Later, it is retired, leaving the surface sensor free once more. The car seat then remained unoccupied for 10 s more. This cycle exposed was repeated 4 times for every case.

3. Results and Discussion

Three cases are being evaluated: people, a heavy object and a bottle of water. Figure 3 shows the experimental results for two people.

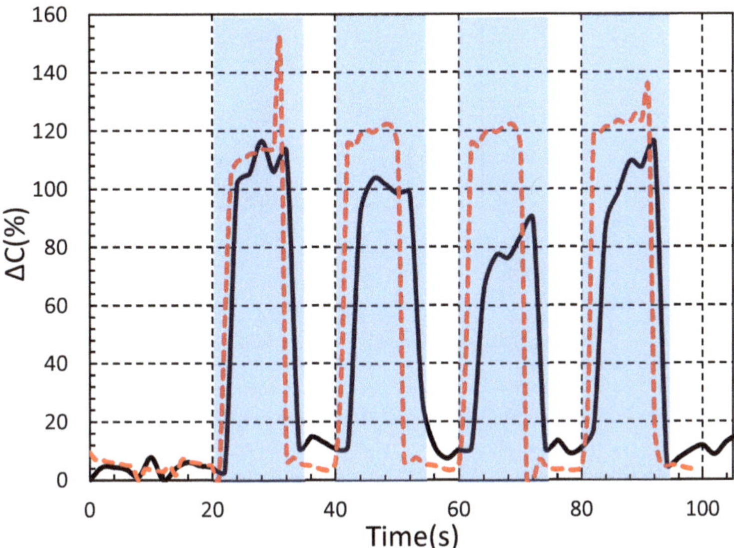

Figure 3. Presence test cycles for two different people: Person 1 (continuous black line); Person 2 (dashed red line); detection zones (blue shaded zone).

As can be seen, both cases clearly differentiate between the occupied (blue zones) and void states. The void state values are around 10% of the capacitance variation, which imply a slight increase from the start of the void state. This increase was due to the humidity or other substances retained by the fabric after each cycle was performed by the person.

Meanwhile, when a person was sat over the sensor, the capacitance variation experienced a significant increase, reaching values of around 100% of the capacitance variation. Some differences in the graph between both people exist. These changes could be related to physical characteristics of the human body which could affect the permittivity, but could also be related to the positioning of the buttock over the sensor. The last situation was observed on the third pulse of Person 1 (continuous line), where the capacitance variation was lower due to the misplacement of their body.

Finally, a test with object was performed. A bag with a weight of 20 kg and a 2 L bottle of water were prepared to be measured. The results are presented in Figure 4. It is important to highlight that the maximum value obtained by the sensor when an object is detected is 32% of the full capacitance variation.

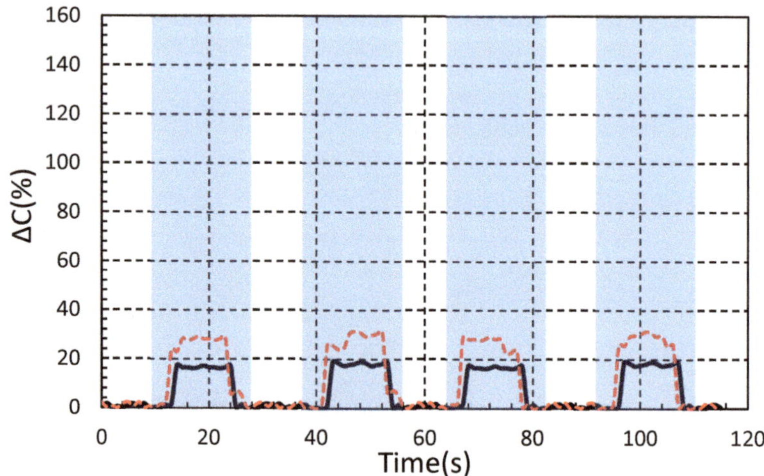

Figure 4. Presence sensor response against objects: 2 L of water (continuous black line); heavy bag (20 kg) (dashed red line); detection zones (blue shaded zone).

The cycle was performed for the 20 kg bag, as can be seen represented by a dashed line in the graph, but the sensor does not show any significant response compared with the person detection. The 20 kg bag was detected with a maximum value of 31% of capacitance variation. The change in capacitance was related with the permittivity difference between the sensor fabric and bag fabric. For the case of the bottle of water, depicted by a continuous line, the values observed demonstrate that the sensor was able to tell when the bottle was placed. The water inside the bottle provoked a change in the surrounding permittivity that resulted in 20% of the full capacitance variation being recorded. As seen in both cases, the weight of the object measured is not relevant for capacitance variation because the capacitance is only affected by permittivity.

In the case of sensing a person's body, it can be observed how, during the void state, the sensor does not return to the initial value due to the remaining humidity from the person's body; meanwhile, for the case of the object, this effect is not observed.

People measurement demonstrates the functionality of the presence capacitance sensor to distinguish between occupied and void situations. The response for a person is clearly differentiable from an object, even when a bottle of water is placed over the sensor. Therefore, this sensor could solve the false positives provoked by the FSR sensor in automotive applications, providing a new way of measurement directly integrated into the upholstery.

4. Conclusions

An automotive seat occupancy sensor based on e-textile technology is presented in this work. The sensor, based on a woven capacitive sensor, solves the problem suffered by FSR sensors. The sensor is capable of differentiating a person from an object due to the permittivity differences between them. The sensor could also be integrated directly into the upholstery of the car seat, saving space where the FSR sensors would otherwise be installed.

Author Contributions: Conceptualization, M.M.-E. and R.F.-G.; methodology, M.M.-E.; validation, M.M.-E., R.F.-G.; formal analysis, M.M.-E., I.G. and R.F.-G.; investigation M.M.-E.; data curation, M.M.-E.; writing—original draft preparation, M.M.-E.; writing—review and editing, M.M.-E., I.G. and R.F.-G.; supervision, I.G. and R.F.-G.; project administration, I.G. and R.F.-G.; funding acquisition, I.G. and R.F.-G. All authors have read and agreed to the published version of the manuscript.

Funding: This research was funded by Spanish Government "Ministerio de ciencia e innovacion" PID2021 124288OB I00 and AGAUR 2020FI-B 00028.

Institutional Review Board Statement: Not applicable.

Informed Consent Statement: Not applicable.

Data Availability Statement: The data presented in this study are available within the article and there is presented in every graph. There is no more data apart from the presented.

Conflicts of Interest: The authors declare no conflict of interest.

References

1. Interlink Electronics Manufactures. Force Sensing Resistors An Overview of the Technology. p. 25. Available online: https://www.sparkfun.com/datasheets/Sensors/Pressure/fsrguide.pdf (accessed on 15 November 2022).
2. Anwary, A.R.; Vassallo, M.; Bouchachia, H. Monitoring of Prolonged and Asymmetrical Posture to Improve Sitting Behavior. In Proceedings of the 2020 International Conference on Data Analytics for Business and Industry: Way Towards a Sustainable Economy (ICDABI), Sakheer, Bahrain, 26–27 October 2020. [CrossRef]
3. Zeeman, A.S.; Booysen, M.J.; Ruggeri, G.; Lagana, B. Capacitive seat sensors for multiple occupancy detection using a low-cost setup. In Proceedings of the 2013 IEEE International Conference on Industrial Technology (ICIT), Cape Town, South Africa, 25–28 February 2013; pp. 1228–1233. [CrossRef]
4. Marc, M.E.; Ignacio, G.; Raul, F.G. A Smart Textile System to Detect Urine Leakage. *IEEE Sens. J.* **2021**, *21*, 26234–26242. [CrossRef]

Proceeding Paper

The Energy Harvesting Performance of a Flexible Triboelectric-Based Electrospun PTFE/PVDF Fibre †

Pattarinee White [1,*], Dmitry Bavykin [2], Mohamed Moshrefi-Torbati [3] and Stephen Beeby [1]

1 Centre for Flexible Electronics and E-Textiles, University of Southampton, Southampton SO17 1BJ, UK
2 Energy Technology Research Group, University of Southampton, Southampton SO17 1BJ, UK
3 Mechatronics Research Group, University of Southampton, Southampton SO17 1BJ, UK
* Correspondence: p.white@soton.ac.uk

† Presented at the 4th International Conference on the Challenges, Opportunities, Innovations and Applications in Electronic Textiles, Nottingham, UK, 8–10 November 2022.

Abstract: A triboelectric power generator/energy harvester is an attractive option for mechanical energy harvesting for smart, wearable applications. This paper reports on the fabrication and evaluation of the energy harvesting performance of Polytetrafluoroethylene/Polyvinylidene Fluoride (PTFE/PVDF) fibre prepared using a one-step electrospinning technique. Different concentrations (0, 1, 2, 3, and 4%wt.) of the 1 μm PTFE powder in the electrospun PVDF fibre were investigated. The electrospun fibre was assembled into a nonwoven fabric mat and tested in the vertical contact separation triboelectric mode by constructing a sandwich structure with electrodes in a book-shaped assembly. The voltage output from the cyclical compressive test for fibres with 4%wt. PTFE in PVDF was five times greater than it was for the 100% PVDF electrospun fibres. The influence of adding nylon fabric as a triboelectric donor material within the assembly was explored. The output of the 4%wt. PTFE/PVDF sample was then tested with and without nylon fabric at different frequencies (3–12 Hz). The results show a further 80% increase in the output voltage with the additional nylon fabric included, and the harvester was able to illuminate up to 95 LEDs.

Keywords: triboelectric; textile energy harvester; electrospinning; electrospun fibre

Citation: White, P.; Bavykin, D.; Moshrefi-Torbati, M.; Beeby, S. The Energy Harvesting Performance of a Flexible Triboelectric-Based Electrospun PTFE/PVDF Fibre. *Eng. Proc.* **2023**, *30*, 8. https://doi.org/10.3390/engproc2023030008

Academic Editors: Kai Yang, Russel Torah and Theodore Hughes-Riley

Published: 20 January 2023

1. Introduction

Polytetrafluoroethylene (PTFE) has good thermal stability, and is the most negative triboelectric material listed in the triboelectric series that are used for energy harvesting applications [1]. Increasing the surface area of the PTFE within a triboelectric harvester is one approach to improve the output of the harvester. Previous attempts to prepare PTFE fibre mats using the electrospinning process involve the use of a precursor polymer solution in order to structure the shape of the PTFE. This requires a subsequent thermal treatment to eliminate the precursor polymer and, if it is used in an electrostatic device or to enhance the triboelectric generation, a corona charging process to restore the trapped charge in the PTFE fibre. The PTFE fibres produced in this manner show good energy harvesting performances, but they require two or more steps in the material preparation [2–5]. Electrospun Polyvinylidene Fluoride (PVDF) fibre has been recognised as a high-performance piezoelectric polymer [6], which also exhibits negative triboelectric properties [1]. PTFE and PVDF can be combined during electrospinning, and this avoids the requirement for the precursor polymer and the subsequent additional processes. The use of electrostatically charged electrospun PTFE/PVDF has been demonstrated in air filtration [7], but its energy harvesting performance has not yet been reported.

The combined PTFE/PVDF fibres can be readily assembled into a non-woven textile, allowing the material to be used in various triboelectric operating modes, such as the lateral sliding mode demonstrated by Paosangthong et al. [8]. The fabrication and testing of the energy harvesting performance of PTFE/PVDF fibre is reported.

2. Materials and Methods

2.1. Electrospinning Preparation

The PTFE/PVDF electrospun fibre was produced by mixing PTFE particles (Sigma Aldrich, Dorset, UK, 1 micron particles) in an 18%wt PVDF solution as a polymer carrier for PTFE. The PVDF solution was prepared from PVDF powder (Sigma Aldrich, Dorset, UK, Mw = 534,000) mixed with N-Dimethylformamide (DMF, Sigma Aldrich, Dorset, UK, 99.8%) and acetone (Fisher Scientific, Waltham, MA, USA, 99.6%) at a 7:3 ratio and mixed on a magnetic hotplate stirrer at 60 °C for 4 h. Different concentrations (0, 1, 2, 3, and 4%wt.) of the PTFE particles were added to the PVDF solvent solution and blended using the magnetic stirrer at room temperature for 2 h. The electrospinning process was performed using a blunt tip (21 G) needle. The electrospinning apparatus EC-DIG produced by IME Technologies, Netherlands was used in this study. The distance from the tip to the substrate, the applied voltage, the flow rate, and the rotating drum speed were kept constant at 22 cm, 25 kV, 2 mL/hr, and 150 rpm, respectively, for each concentration. After electrospinning for 90 min, the PTFE/PVDF fibre was collected in the form of a nonwoven fibre mat, as shown in Figure 1a. The energy harvesting performance of the electrospun PTFE/PVDF fibre mats were measured without any further processing steps.

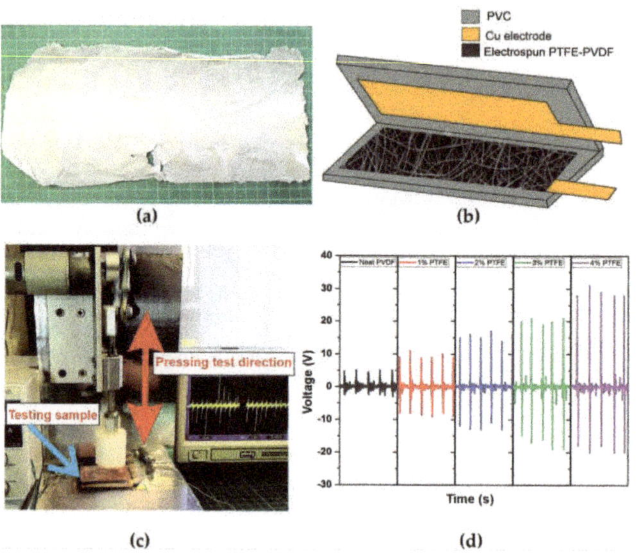

Figure 1. (**a**) PTFE/PVDF fibre mats after the electrospinning process of the 4%wt PTFE in the PVDF sample; (**b**) the schematic of the booked shape assembly for the testing cell PTFE/PVDF fibre sandwich with electrodes; (**c**) the compression test rig; (**d**) voltage output from constantly tapping the booked shape PTFE/PVDF fibre assembly at different PTFE concentrations of 0, 1, 2, 3, and 4%wt., respectively.

2.2. Test Assembly and Protocol

The electrospun PTFE/PVDF fibre mats with each concentration of PTFE (0, 1%, 2%, 3%, and 4%) were cut into 5 × 4 cm samples. Each sample was then assembled in a sandwich structure with Cu electrodes using a folded over (book-shaped) PVC sheet backing, as shown in Figure 1b. This forms a vertical contact separation mode triboelectric harvester. A second generator design with a piece of nylon fabric added to the assembly as a triboelectric donor material was used to explore the addition of this material for the enhancement of the performance of the triboelectric power generator.

A cyclical compression test system using a linear actuator was set up to apply a controlled pressure of 0.5 N/cm^2 at different frequencies (3–12 Hz), as shown in Figure 1c.

The test sample was attached to the oscilloscope to observe the changes in the voltage induced by the periodic mechanical pressure applied by the rig. The capacitor charging experiment was performed using a full-wave bridge rectifier to charge 0.1, 1, 10, and 100 μF capacitors.

3. Results and Discussion

The vertical book-shaped structure was chosen for energy harvesting performance testing as it is a simple structure and could be assembled into a multi-layered device. The highest voltage output from compressing at 5 Hz was found to be 30 V for the 4%wt PTFE, which is five times higher than that of the 100% PVDF sample (6 V), as shown in Figure 1d. This clearly shows that introducing the PTFE particles in the PVDF fibre can improve the performance of triboelectric power generators. The output voltage increases with an increasing percentage of PTFE particles. However, 4% PTFE is the highest amount of polymer content that can be processed via electrospinning because the solution conductivity is not strong enough to produce the electrospun fibre.

The 4% PTFE in the PTFE/PVDF sample was used in the energy harvesting test at different pressing rates of 3–12 Hz. It was found that the output voltage increases with an increasing pressing frequency. The highest output voltage of around 55 V was found at 12 Hz, as shown in Figure 2b. To enhance the energy harvesting performance of the triboelectric power generator, nylon fabric, which exhibits a high positive affinity in the triboelectric material series, was placed on top of the top copper electrode as shown in Figure 2a. The voltage output of the Nylon-PTFE/PVDF device at the different pressing frequencies is shown in Figure 2c. A small improvement in the voltage output was observed with the highest voltage output, which reached at 70 V at 12 Hz.

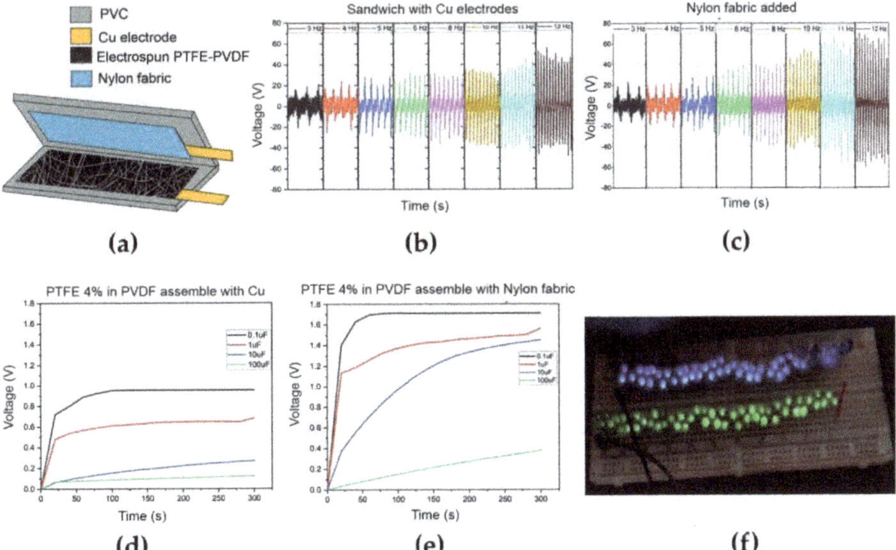

Figure 2. (**a**) The schematic of the improved triboelectric harvester design with added nylon fabric, (**b**) the voltage output of the triboelectric assembly without nylon fabric, (**c**) the voltage output of the triboelectric assembly after introducing nylon fabric, (**d**) the charging profile, voltage vs. time of the PTFE/PVDF fibre device without nylon fabric, (**e**) the charging profile after introducing nylon fabric and (**f**) 95 LED lights were illuminated via tapping the harvester with the PTFE/PVDF electrospun fibres as the acceptor and nylon fabric as the donor material.

The charging experiment was further performed to explore and compare the amounts of energy captured by the device and transferred to the capacitor storage. Devices with and without the nylon fabric were tested at 5 Hz with the different capacitor values, and the results are shown in Figure 2d,e. Overall, the energy stored in the capacitors is higher for the nylon-PTFE/PVDF device, with the amounts of energy stored being 0.15, 1.0, 7.2 and 4.5 μJ for 0.1, 1, 10 and 100 μF, respectively. The maximum energy stored value occurs with the 10 μF capacitor, which nearly reached its maximum capacity in the 300 s charging time. This is 69%, 79%, 94% and 89% larger than those of the device without nylon for 0.1, 1, 10 and 100 μF, respectively. The energy stored in the 100 μF capacitor is less than that of the 10 μF capacitor over the same duration due to the impedance mismatch.

After connecting the optimum harvester structure (4% PTFE with Nylon fabric) to a full-wave bridge rectifier circuit, it was found that 95 LED lights were illuminated when it was compressed at 12 Hz. The charging experiment and illuminating LED light results demonstrate the promising mechanical energy conversion that was achieved with the electrospun PTFE/PVDF fibre mat. The lightweight, flexible, and breathable PTFE/PVDF electrospun fibre could be integrated within clothing as an energy source, whilst remaining comfortable for the user.

4. Conclusions

A flexible Polytetrafluoroethylene (PTFE) fibre was successfully prepared using a one-step electrospinning process using a Polyvinylidene Fluoride (PVDF) solution as a precursor. The energy harvesting performance was first explored using a vertical contact separation mode triboelectric assembly. The electrospun fibres were collected in the form of a non-woven textile mat, which displayed a very promising negative surface potential. The voltage output was increased by a factor of five by adding PTFE to the PVDF electrospun fibre. This was further improved by the addition of a nylon fabric with the triboelectric harvester. The textile harvester was shown to illuminate up to 95 LED lights when it was assembled with nylon fabric as a donor material.

Author Contributions: Conceptualization, P.W. and S.B.; methodology, P.W.; formal analysis, P.W.; investigation, P.W.; resources, D.B., M.M.-T. and S.B.; data curation, P.W.; writing—original draft preparation, P.W., M.M.-T., D.B. and S.B.; writing—review and editing, P.W., M.M.-T., D.B. and S.B.; visualization, P.W.; supervision, M.M.-T., D.B. and S.B.; project administration, S.B.; funding acquisition, S.B. All authors have read and agreed to the published version of the manuscript.

Funding: This research received no external funding.

Institutional Review Board Statement: Not applicable.

Informed Consent Statement: Not applicable.

Data Availability Statement: The data presented in this study are available upon request from the corresponding author.

Acknowledgments: Pattarinee White would like to thank the Royal Thai Government Scholarship and Rajamangala University of Technology Krungthep for kindly supporting this research. The work of Stephen Beeby was supported by the U.K. Royal Academy of Engineering under the Chairs in Emerging Technologies Scheme.

Conflicts of Interest: The authors declare no conflict of interest.

References

1. Diaz, A.F.; Felix-Navarro, R.M. A Semi-Quantitative Tribo-Electric Series for Polymeric Materials: The Influence of Chemical Structure and Properties. *J. Electrost.* **2004**, *62*, 277–290. [CrossRef]
2. Su, C.; Li, Y.; Cao, H.; Lu, C.; Li, Y.; Chang, J.; Duan, F. Novel PTFE Hollow Fiber Membrane Fabricated by Emulsion Electrospinning and Sintering for Membrane Distillation. *J. Membr. Sci.* **2019**, *583*, 200–208. [CrossRef]
3. Lin, S.; Cheng, Y.; Mo, X.; Chen, S.; Xu, Z.; Zhou, B.; Zhou, H.; Hu, B.; Zhou, J. Electrospun Polytetrafluoroethylene Nanofibrous Membrane for High-Performance Self-Powered Sensors. *Nanoscale Res. Lett.* **2019**, *14*, 251. [CrossRef] [PubMed]
4. Zhao, P.; Soin, N.; Prashanthi, K.; Chen, J.; Dong, S.; Zhou, E.; Zhu, Z.; Narasimulu, A.A.; Montemagno, C.D.; Yu, L.; et al. Emulsion Electrospinning of Polytetrafluoroethylene (PTFE) Nanofibrous Membranes for High-Performance Triboelectric Nanogenerators. *ACS Appl. Mater. Interfaces* **2018**, *10*, 5880–5891. [CrossRef] [PubMed]

5. Lu, Y.; Amroun, D.; Leprince-Wang, Y.; Basset, P. A Paper-Based Electrostatic Kinetic Energy Harvester with Stacked Multiple Electret Films Made of Electrospun Polymer Nanofibers. *J. Phys. Conf. Ser.* **2016**, *773*, 012032. [CrossRef]
6. Lam, T.N.; Wang, C.C.; Ko, W.C.; Wu, J.M.; Lai, S.N.; Chuang, W.T.; Su, C.J.; Ma, C.Y.; Luo, M.Y.; Wang, Y.J.; et al. Tuning Mechanical Properties of Electrospun Piezoelectric Nanofibers by Heat Treatment. *Materialia* **2019**, *8*, 100461. [CrossRef]
7. Dong, Z.Q.; Ma, X.H.; Xu, Z.L.; You, W.T.; Li, F.B. Superhydrophobic PVDF-PTFE Electrospun Nanofibrous Membranes for Desalination by Vacuum Membrane Distillation. *Desalination* **2014**, *347*, 175–183. [CrossRef]
8. Paosangthong, W.; Wagih, M.; Torah, R.; Beeby, S. Textile-Based Triboelectric Nanogenerator with Alternating Positive and Negative Freestanding Grating Structure. *Nano Energy* **2019**, *66*, 104148. [CrossRef]

Proceeding Paper

Second Skins: Exploring the Challenges and Opportunities for Designing Adaptable Garments Using E-Textile †

Malou Beemer [1,2,*], Christian Dils [3] and Troy Nachtigall [2,4]

1 Atelier Mlou, 5921 AV Venlo, The Netherlands
2 Wearable Data Studio, Amsterdam University of Applied Sciences, 1091 GC Amsterdam, The Netherlands
3 Institute for Reliability and Microintegration, Fraunhofer IZM, 13355 Berlin, Germany
4 Industrial Design, Eindhoven University of Technology, 5612 AZ Eindhoven, The Netherlands
* Correspondence: m.beemer@hva.nl; Tel.: +31-6-24608283
† Presented at the 4th International Conference on the Challenges, Opportunities, Innovations and Applications in Electronic Textiles, Nottingham, UK, 8–10 November 2022.

Abstract: The process of making adaptive and responsive wearables on the scale of the body has often been a process where designers use off-the-shelf parts or hand-crafted electronics to fabricate garments. However, recent research has shown the importance of emergence in the process of making. *Second Skins* is a multistakeholder exploration into the creation of those garments where the designers and engineers work together throughout the design process so that opportunities and challenges emerge with all stakeholders present in the process. This research serves as a case study into the creation of adaptive caring garments for sustainable wardrobes from a multistakeholder design team. The team created a garment that can customize the colors, patterns, structures, and other properties dynamically. A reflection on the multi-stakeholder process unpacks the process to explore the challenges and opportunities in adaptable e-textiles.

Keywords: dynamic garments; multistakeholder design; textiles; e-textiles; fabrication

Citation: Beemer, M.; Dils, C.; Nachtigall, T. Second Skins: Exploring the Challenges and Opportunities for Designing Adaptable Garments Using E-Textile. *Eng. Proc.* **2023**, *30*, 9. https://doi.org/10.3390/engproc2023030009

Academic Editors: Steve Beeby, Kai Yang, Russel Torah and Theodore Hughes-Riley

Published: 20 January 2023

1. Introduction

Recent research in computer–human interaction using research through design [1] has highlighted the importance of the ideas that emerge [2,3] as fundamental to the design considerations that come together to create the final designed item. Often in the design of dynamic garments, the stakeholders who create the electronics are separate from the stakeholders who design the garment, unless low-level prototyping is taking place [4–6]. The design process that emerges has been shown to be detrimental to the outcomes of a project, yet rarely understood or studied [7]. In this paper, we present our process of formulating and realizing a dynamic garment where technological and aesthetic design considerations are undertaken together in an emergent process. This includes the ideation of the garment, a modular design approach, and whether to visualize a reactive light as a material or the utilization of soft actuators. The authors, as makers, reflect upon the design considerations of the design process and how the final design supports and cares for the wearer, not only on a practical but also on an emotional and social level. The created prototype serves not only as a functional forerunner, but also as a showroom prototype [8], inviting dialog about the process to discuss the trajectories of the research, how it could have been undertaken, and how that informs design spaces still waiting to be explored.

2. Design Process

This research was undertaken in the context of the Re-FREAM pillar of the European STARTS Programme. The project was a collaborative and interdisciplinary research effort that included stakeholders Malou Beemer, Fraunhofer IZM, Profactor, EMPA, and Wear It Berlin [9]. The team worked for nine months to co-create the project, including three months in the technology partners' labs. Due to the COVID-19 pandemic, the process was mostly online which may have influenced our choice of material and technique used in creating adaptive garment prototypes. Yet at the same time, it enabled new collaboration thanks to the availability of the partners.

2.1. Soft Actuators

Inspired by numerous soft actuator projects that are integrated into textile-based objects by international research teams [9–13], such as the MIT Tangible Media Group or Harvard BioDesign Lab, an intensive literature study was carried out. Several relevant scientific articles were studied, and expert interviews were conducted, including with the Fraunhofer ISC Smart Materials Centre, MIT, and the RCA Soft System Research Group. As a result, five types of actuation methods were further investigated. Shape Memory Alloys, including Nickel Titanium, were rejected because they were not innovative enough and require high power consumption. Pneumatic artificial muscles were not yet considered wearable or convenient due to the pumps and valves required. IPMCs were excluded because samples could not be produced and access to third-party material was not available. Electric fields that triggered soft actuators had a risk of injury to the human body due to high currents and voltages. A new concept, consisting of textile/polymer laminate that deforms due to different coefficients of thermal expansion, was attempted but experiments on this were not successful. An overview of the literature research is presented online [14]. These explorations brought the team to the conclusion that actuators remain overly challenging at this time. Therefore, we decided to focus on using adaptive light integration for the prototypes.

2.2. Modular Design Approach

From the early explorations, we started to explore a modular design approach, as depicted in Figure 1, that leads to new dynamics, interactions, and business models. Taking the disassembly and the end of use into consideration during the design process takes extra effort and brings a lot of challenges, from construction and connection methods to material selection and manufacturing. This was even more essential when we integrate electronics and other technologies into garments, fusing two polluting industries with different waste streams. Our design approach brought a new wave of aesthetics with it that will not only elevate garment designs on a sustainable level but can also evolve into a new language regarding the shape, structure, and overall feel. These ideas were confirmed by our findings when we were exploring the Planetary Design Tool by Max Marwede and Robin Hoske from Fraunhofer IZM [15].

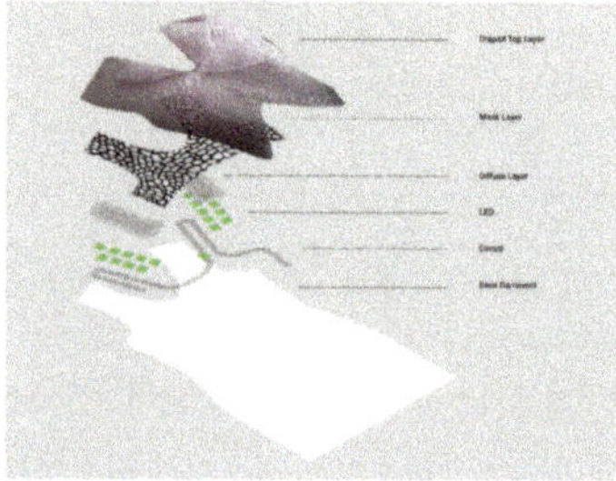

Figure 1. Overview of the garment layers devised for a modular system.

For the final prototype, we developed two undergarments with integrated modules from the Fraunhofer IZM hardware kit for e-textiles [16]. Embroidered conductor tracks, made of insulated conductive thread, connect the circuit boards with LEDs and sensors to the main control via the e-textile-bonding technique [17]. During our user test event, we learned that an input is required to be able to easily change the pre-programmed light or color patterns and to be able to switch the light on and off. This was fully in line with the concept of adapting to how much you want to attract attention or become less noticeable. We, therefore, use the data from the Inertial Measurement Unit to implement a tap sensor, as displayed in Figure 2. A light-diffusing layer is added to create a soft skin-like light effect, which is covered by an interchangeable mask layer to create different patterns as shadow play in the garment.

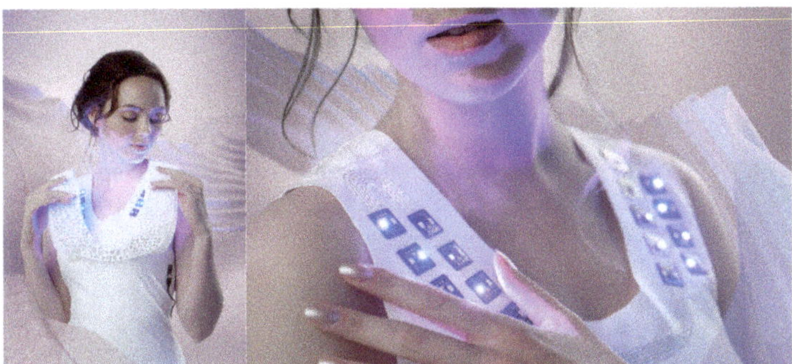

Figure 2. Diffuse and mask layer with parametric patterns (**left**) and Undergarment with Embroidered conductor tracks and bonded PCBs (**right**).

2.3. Outergarments

The top garments are a fusion of old craftsmanship and new digital technologies. The ombre color effect is created with sublimation printing on laser-cut textile elements. To create a bigger diffuse effect we used a hand pleating technique to develop three-dimensional textures in the fabric, as shown in Figure 3.

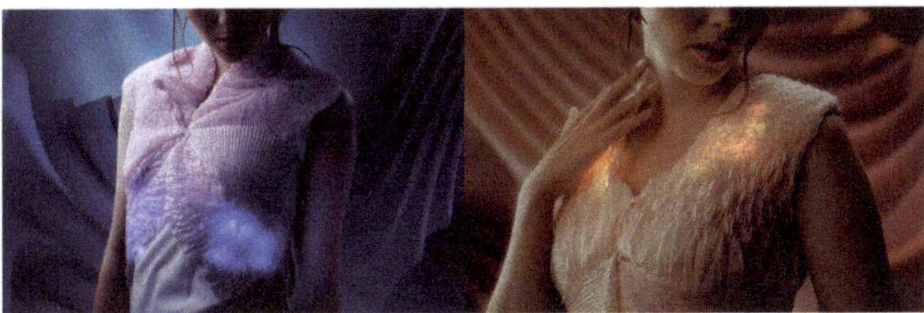

Figure 3. Pleated top garment used as a dynamic projection surface.

3. Reflection and Discussion

3.1. The Gap between Research and Reality

The biggest hurdle in the project was finding a fitting technique and material that responded to the design considerations of the organic change we designed. In the areas of wearable tech, e-textiles, and soft actuators, there is a large gap between existing research and actual technology implementation. Scientific research is often done within the parameters of a lab with specific conditions, which is an achievement. However, multistakeholder exploration of a fluid translation and collaborative connection between the technology and the application is where we can make a big impact.

3.2. The Evolution of Fashion Tech

For centuries, high fashion has been accepted for its aesthetic and expressive value. Fashion tech is a cross-pollination of this field with innovative technologies. The industry of technology is function-driven and often shuns the use of new materials and technological possibilities as insufficient. Instead, the integration of tech in fashion is justified by its purpose and function, which often leads to integration regarding performance improvement, health, comfort, and safety [18]. We take the next step and create a *caring garment* that goes beyond straightforward functionality into expression and communication. Over the past decade, we developed wearables mainly from the tech perspective with the purpose of solving practical issues. Envision a future where expression and functionality seamlessly merch together. With this project, we demonstrate how stakeholder collaboration from the beginning enables dynamics and aesthetics that go hand in hand.

3.3. The Time Frames of Wearable Innovation

Project partner Thomas Gnahm from Wear it Berlin said it best: innovation takes time [19]. Fashion tech is still in the early days, Gnahm compares it to the development of cars. Previously, the first cars looked like carriages without horses. Once we saw the potential of new materials, manufacturing and even forms of math were created to facilitate the process. Fashion tech is still in an age of experimentation and definition. Yet the knowledge required is often too complex for a single person and requires teams that, together, are good at exploration, playfulness, experimentation, and definition.

3.4. Sustainability Positioning

When we talk about sustainability, there is a lot of focus on material sourcing and manufacturing [20]. This is of course essential; however, we believe we tend to neglect the other parts of the value chain that are equally important. We can shift to using better materials and production processes, but this will remain to be an issue if we do not change our behavior and relationship towards our "properties". If we keep consuming and trashing, it will still be a linear process. If we choose to "own" a product, we also need to take responsibility and take ownership of it, by investing in a longer-term valuable relationship. This is not only up to the wearers themselves, but it is the task of designers, producers, and suppliers to guide consumers into this new headspace and is important to regard this during the design process.

3.5. Sustainability and Creativity Balance and Context

Rapid change is currently occurring in the fashion and textile industry. New EU sustainability regulations demand new design and manufacturing methods, as well as new technological developments that create new possibilities for creation. This asks for a wave of new-generation designers that are trained and literate in new digital technologies and sustainable thinking. While, at the same time, taking the heritage and knowledge of traditional textile craftsmanship into regard. Finding a healthy, productive, and innovative balance between philosophical questioning and existential pragmatic thinking within the design domain is key, keeping the bigger sustainable picture in mind while not losing the aesthetic values we create in design practice.

Author Contributions: Conceptualization, methodology, resources, M.B.; investigation, M.B., C.D.; writing—original draft preparation, writing—review and editing, M.B., T.N., C.D. All authors have read and agreed to the published version of the manuscript.

Funding: This research was funded via the S-T-ARTS Re-FREAM project by the European Union Horizon 2020 research and innovation program under Grant Agreement No. 825647.

Institutional Review Board Statement: Not applicable.

Informed Consent Statement: Not applicable.

Data Availability Statement: Not applicable.

Acknowledgments: The authors would like to thank Sigrid Rotzler, Pauline Stockmann, and Janin-Anne Schulze for the embroidery of the conductive patterns; Kamil Garbacz for developing and programming the hardware modules; Lars Stagun for bonding the modules onto the textile; and Martin Haubenreisser for laser cutting the designs. Marina Toeters for support and facilities at Fashion Tech Farm Eindhoven, The Netherlands. Julia van Zilt for technical support and (parametric) pattern creation.

Conflicts of Interest: The authors declare no conflict of interest. The funders had no role in the design of the study; in the collection, analyses, or interpretation of data; in the writing of the manuscript; or in the decision to publish the results.

References

1. Hauser, S.; Oogjes, D.; Wakkary, R.; Verbeek, P.P. An annotated portfolio on doing postphenomenology through research products. In Proceedings of the 2018 Designing Interactive Systems Conference, Hong Kong, China, 9–13 June 2018; pp. 459–472. [CrossRef]
2. Wakkary, R.; Maestri, L. Aspects of Everyday Design: Resourcefulness, Adaptation, and Emergence. *Int. J. Hum. Comput. Interact.* **2008**, *24*, 478–491. [CrossRef]
3. Gaver, W.W.; Krogh, P.G.; Boucher, A.; Chatting, D. Emergence as a Feature of Practice-based Design Research. In Proceedings of the 2022 ACM Designing Interactive Systems Conference: Digital Wellbeing, Virtual, 13–17 June 2022; pp. 517–526. [CrossRef]
4. Frankjaer, R.; Dalsgaard, P. Understanding Craft-Based Inquiry in HCI. In Proceedings of the 2018 on Designing Interactive Systems Conference, Hong Kong, China, 9–13 June 2018; pp. 473–484. [CrossRef]
5. Devendorf, L.; Rosner, D.K. Beyond Hybrids. In Proceedings of the 2017 Conference on Designing Interactive Systems, Edinburgh, UK, 10–14 June 2017; pp. 995–1000. [CrossRef]
6. Berzowska, J.; Kelliher, A.; Rosner, D.K.; Ratto, M.; Kite, S. Critical Materiality. In Proceedings of the Thirteenth International Conference on Tangible, Embedded, and Embodied Interaction, Tempe, AZ, USA, 17–20 March 2019; pp. 691–694. [CrossRef]
7. da Rocha, B.G.; Andersen, K.; Tomico, O. Portfolio of Loose Ends. In Proceedings of the 2022 ACM Designing Interactive Systems Conference: Digital Wellbeing, Virtual, 13–17 June 2022; pp. 527–540. [CrossRef]
8. Matthews, B.; Wensveen, S. Prototypes and prototyping in design research. In *Routledge Companion to Design Research*; Paul, A., Rodgers, J.Y., Eds.; Routledge: London, UK, 2014; pp. 262–276. Available online: https://www.researchgate.net/publication/270511639 (accessed on 30 June 2019).
9. Thalman, C.; Artemiadis, P. A review of soft wearable robots that provide active assistance: Trends, common actuation methods, fabrication, and applications. *Wearable Technol.* **2020**, *1*, e3. [CrossRef]
10. Li, M.; Pal, A.; Aghakhani, A.; Pena-Francesch, A.; Sitti, M. Soft actuators for real-world applications. *Nat. Rev. Mater.* **2021**, *7*, 235–249. [CrossRef] [PubMed]
11. Kilic Afsar, O.; Shtarbanov, A.; Mor, H.; Nakagaki, K.; Forman, J.; Modrei, K.; Jeong, S.H.; Hjort, K.; Höök, K.; Ishii, H. OmniFiber: Integrated Fluidic Fiber Actuators for Weaving Movement based Interactions into the Fabric of Everyday Life. In Proceedings of the 34th Annual ACM Symposium on User Interface Software and Technology, Virtual, 10–14 October 2021; pp. 1010–1026. [CrossRef]
12. Sanchez, V.; Payne, C.J.; Preston, D.J.; Alvarez, J.T.; Weaver, J.C.; Atalay, A.; Boyvat, M.; Vogt, D.M.; Wood, R.J.; Whitesides, G.M.; et al. Walsh. Smart Thermally Actuating Textiles. *Adv. Mater. Technol.* **2020**, *5*, 2000383. [CrossRef]
13. Connolly, F.; Wagner, D.A.; Walsh, C.J.; Bertoldi, K. Sew-free anisotropic textile composites for rapid design and manufacturing of soft wearable robots. *Extreme Mech. Lett.* **2019**, *27*, 52–58. [CrossRef]
14. Second Skins Research on Responsive and Adaptive Materials. Available online: https://re-fream.eu/second-skins-research-on-responsive-and-adaptive-materials/ (accessed on 12 December 2022).
15. Planetary Design Circle—A Holistic and Strategic Design Tool—Re-FREAM. Available online: https://re-fream.eu/planetary-design-circle-a-holistic-and-strategic-design-tool/ (accessed on 12 December 2022).
16. Garbacz, K.; Stagun, L.; Rotzler, S.; Semenec, M.; von Krshiwoblozki, M. Modular E-Textile Toolkit for Prototyping and Manufacturing. *Proceedings* **2021**, *68*, 5. [CrossRef]
17. von Krshiwoblozki, M.; Linz, T.; Neudeck, A.; Kallmayer, C. Electronics in Textiles-Adhesive Bonding Technology for reliably embedding Electronic Modules into Textile Circuits. *Adv. Sci. Technol.* **2012**, *85*, 1–10. [CrossRef]
18. Toeters, M. *Unfolding Fashion Tech: Pioneers of Bright Futures 2000–2020*; Onomatopee: Eindhoven, The Netherlands, 2019.
19. Gnahm, T.; Wear it GmbH, Berlin, Germany. Personal communication, 2021.
20. Eppinger, E.; Slomkowski, A.; Behrendt, T.; Rotzler, S.; Marwede, M. Design for Recycling of E-Textiles: Current Issues of Recycling of Products Combining Electronics and Textiles and Implications for a Circular Design Approach. In *Recycling—Recent Advances*; Saleh, H.M., Hassan, A.I., Eds.; IntechOpen: London, UK, 2022. [CrossRef]

Proceeding Paper

Development of a Knitted Strain Sensor for Health Monitoring Applications †

Beyza Bozali *, Sepideh Ghodrat and Kaspar M. B. Jansen

Faculty of Industrial Design Engineering, Delft University of Technology, 2628 CE Delft, The Netherlands
* Correspondence: b.bozali@tudelft.nl
† Presented at the 4th International Conference on the Challenges, Opportunities, Innovations and Applications in Electronic Textiles, Nottingham, UK, 8–10 November 2022.

Abstract: As an emerging technology, smart textiles have attracted attention for rehabilitation purposes to monitor heart rate, blood pressure, breathing rate, body posture and limb movements. Compared with traditional sensors, knitted sensors constructed from conductive yarns are breathable, stretchable and washable, and therefore, provide more comfort to the body and can be used in everyday life. In this study, knitted strain sensors were produced that are linear with up to 40% strain, sensitivity of 1.19 and hysteresis of 1.2% in absolute values, and hysteresis of 0.03 when scaled to the working range of 40%. The developed sensor was integrated into a wearable wrist-glove system for finger and wrist monitoring. The results show that the wearable was able to detect different finger angles and positions of the wrist.

Keywords: knitted strain sensor; health monitoring applications; smart textiles; wearable textiles

Citation: Bozali, B.; Ghodrat, S.; Jansen, K.M.B. Development of a Knitted Strain Sensor for Health Monitoring Applications. *Eng. Proc.* **2023**, *30*, 10. https://doi.org/10.3390/engproc2023030010

Academic Editors: Steve Beeby, Kai Yang, Russel Torah and Theodore Hughes-Riley

Published: 29 January 2023

1. Introduction

Flexible and wearable sensors have gained attention in recent years for a variety of applications, including human–device interfaces and the monitoring of health indicators such as respiration rate, heart rate and body position [1–3]. Conventional sensors are often integrated into structures as an external element or attached to the surface, but these create discomfort for the user due to the bulky and rigid nature of electronic devices such as IMUs for health monitoring purposes [4]. In this context, textile-based strain sensors offer a new generation of devices that combine wearability, lightness, comfort and stretchability with strain-sensing functionality. They can be comfortably worn and sense a wide range of body strains for a vast number of health monitoring applications, thus making them a good alternative to traditional bulky electronic sensors and making wearable systems more feasible [5].

By using textile-based strain sensors, it is possible to investigate and identify the ideal rehabilitation posture by analyzing the physiological properties of finger and wrist movements. For these wearable sensors, several materials and methods have been investigated to monitor the different parts of the body such as the finger, wrist, arm and leg for rehabilitation purposes. Ryu et al. investigated the performance of the knitted strain sensor in a glove by using silver-plated yarns to distinguish the finger movements, and the electrical responses of the compressive strain demonstrated strong stability and linearity through various finger rolling angles [6]. Lee et al. also concluded that the developed glove might be useful to amputees as a tool that allows them to rehabilitate or regulate a myoelectric prosthesis by putting the sensing elements into the glove and producing the whole-garment knitting technique for ease of commercialization [7]. Isaia et al. evaluated the performance of strain sensors knitted with various conductive yarns in terms of their sensing properties, hysteresis and comfort for joint motion-tracking applications during repetitive flexion–extension cycles [8,9]. Textile-based and knitted strain sensors have been proposed in many studies but have been less useful in practical applications due to the differences in measured strain during loading and unloading (hysteresis). In recent work [10], however, we were able to develop knitted strain sensors with extremely low hysteresis values. In this study, we integrated those sensors into a wrist-to-finger wearable demonstrator, the performance of which was tested.

2. Materials and Methods

In the previous study [10], a textile-based strain sensor was developed with a specific knitting technique, and its electromechanical performance was tested and reported as a hysteresis of 0.03 and a gauge factor of 1.19; in the following study, this newly developed strain sensor was integrated into a wrist-to-finger wearable system. A 1 × 1 rib-knit design was chosen for the strain sensor design and knitted on a Stoll CMS 530 machine using conductive yarn from Shieldex with a yarn count of dtex 235 and initial resistance of $\leq$600 Ω/m, and elastic yarn from Yeoman of Nm 15. Utilizing the plating technique, knitted strain sensors were produced by positioning the conductive yarn inside and the elastic yarn on the outside of the knitted structure (See Figure 1). In the second stage of the study, this developed knitted strain sensor was integrated into the wrist-to-finger wearable system to monitor the movements of the finger and wrist, as shown in Figure 2. Apart from the knitted sensor part in the wearable system, woven cotton fabric was selected for the rest of the design to be able to have a non-elastic and adjustable structure. A Bluetooth Arduino Nano and a power supply were integrated into the wrist-to-finger design using conductive yarns. According to the finger and hand size, the design can be easily modified, and this helps with ease of manufacturing in the later stages. The electromechanical performance of the knitted strain sensor was tested in the course-wise direction by performing four test cycles at 30 mm/min, using a custom-made tensile tester, and the resistance response during tensile extension–relaxation tests was assessed. The performance of the wrist-to-glove design during finger and wrist movements was evaluated and the movements were recorded.

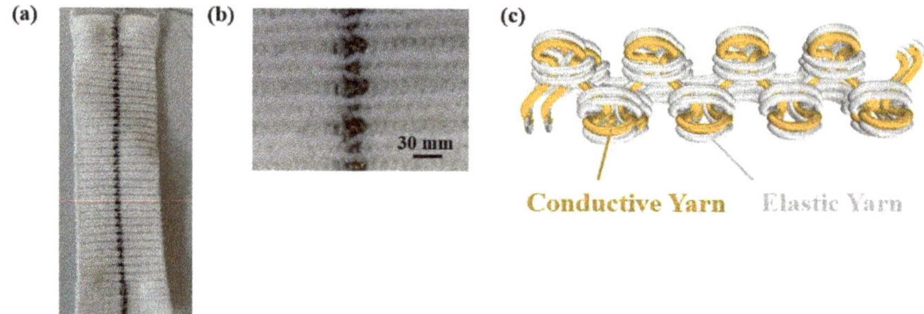

Figure 1. (**a**) The developed knitted strain sensor; (**b**) optical images of the sensing region which shows the conductive yarns (yellow) positioned inside and elastic yarns outside (white); and (**c**) illustration of the conductive and elastic yarn positioning within the knitted structure.

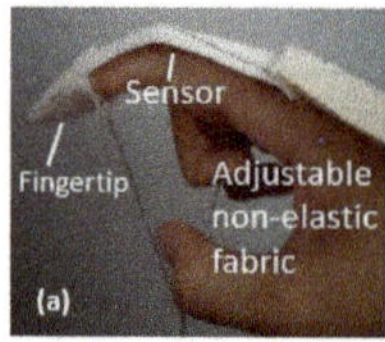

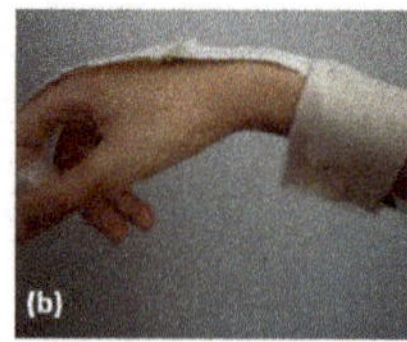

Figure 2. The wearable sensor-glove system with a knitted strain sensor to monitor (**a**) finger movements and (**b**) wrist movement.

3. Results

3.1. Electromechanical Performance of the Knitted Strain Sensor

The electromechanical performance of the knitted strain sensor was investigated and is illustrated in Figure 3. The sensor works linearly over a range of 40% with a hysteresis value and gauge factor of 0.03 and 1.19, respectively. The hysteresis along the strain axis amounts to 1.2% (absolute value) and 0.03 when scaled to the working range of 40% [10]. Because of the high linearity and low hysteresis, this developed sensor is found to be promising for monitoring finger or wrist movement, and the sensor is integrated into the wearable system to test its performance.

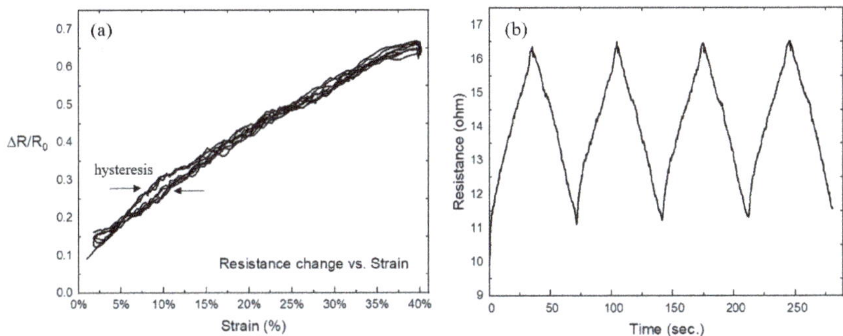

Figure 3. The developed knitted strain sensor graphs under four cyclic tests: (**a**) relative resistance change versus strain and (**b**) resistance versus time.

3.2. Performance Evaluation of Wrist-to-Finger Monitoring System

Bending the finger and wrist deforms the fabric, causing the sensor to change resistance. In this way, finger and wrist movements can be directly detected and monitored. Figure 4a shows the results of finger motion detection at different bending angles by the wrist-to-finger wearable system, targeted at 30°, 60° and 90°. This shows that with the increase in the bending angle of the finger joint from 30° to 60°, and then to 90°, the change in the relative resistance value increases. To investigate the wrist monitoring, the same test was applied to the wrist-to-finger wearable system, and the plot also provides distinguishable patterns in the flexed and unbent positions, as shown in Figure 4b. Note that tests were performed to demonstrate the effects, and the bending angles were not well controlled. However, the observed signal changes were seen to be measurable.

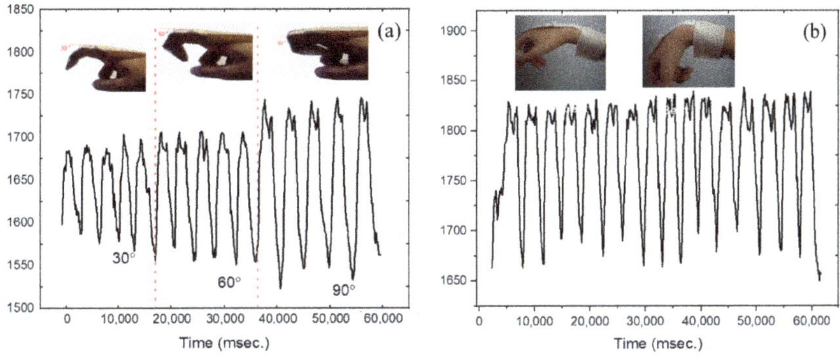

Figure 4. The wrist-to-finger wearable system: (**a**) finger movement monitoring at different angles of 30, 60 and 90°, and (**b**) wrist monitoring.

4. Discussion

For applications such as posture monitoring, VR and rehabilitation, it is desired to have integrated, soft and stretchable strain sensors. We used a novel knitted sensor to construct a wrist-to-finger demonstrator. The measurement showed that it is possible to distinguish between the bending angles and loads of the fingers and the wrist [10]. When the sensor is implemented into the wrist-to-finger system, overall, it performs linearly. The small differences observed in peak values during the tests can be attributed to both the slipping of the garment with respect to the skin and the reproducibility errors inherent to manual movements. In a future study, these irregularities could be amended by fixing the wearable system to the joints of the body.

5. Conclusions

We produced a linear knitted strain sensor with low hysteresis and a working range of at least 40%. The developed knitted sensor can easily be utilized as a part of a wearable system to monitor finger and wrist movement without interfering with the existing fabric performance and appearance. The adjustable wearable system demonstrates the usefulness of the newly developed sensors and has the potential to be used for rehabilitation purposes in health monitoring applications.

Author Contributions: B.B. conceived the study and carried out the testing; S.G. and K.M.B.J. supervised and reviewed the manuscript. All authors have read and agreed to the published version of the manuscript.

Funding: Beyza Bozali was funded by the Turkish Ministry of National Education under the study abroad program for her PhD studies.

Institutional Review Board Statement: Not applicable.

Informed Consent Statement: Not applicable.

Data Availability Statement: Not applicable.

Conflicts of Interest: The authors declare no conflict of interest.

References

1. Zhang, L.; He, J.; Liao, Y.; Zeng, X.; Qiu, N.; Liang, Y.; Xiao, P.; Chen, T. A self-protective, reproducible textile sensor with high performance towards human–machine interactions. *J. Mater. Chem.* **2019**, *7*, 26631–26640. [CrossRef]
2. Li, B.; Xiao, G.; Liu, F.; Qiao, Y.; Li, C.M.; Lu, Z. A flexible humidity sensor based on silk fabrics for human respiration monitoring. *J. Mater. Chem. C* **2018**, *6*, 4549–4554. [CrossRef]
3. Massaroni, C.; Di Tocco, J.; Presti, D.L.; Longo, U.G.; Miccinilli, S.; Sterzi, S.; Formica, D.; Saccomandi, P.; Schena, E. Smart Textile Based on Piezoresistive Sensing Elements for Respiratory Monitoring. *IEEE Sens. J.* **2019**, *19*, 7718–7725. [CrossRef]
4. Lin, B.S.; Lee, I.J.; Yang, S.Y.; Lo, Y.C.; Lee, J.; Chen, J.L. Design of an Inertial-Sensor-Based Data Glove for Hand Function Evaluation. *Sensors* **2018**, *18*, 1545. [CrossRef] [PubMed]
5. Hatamie, A.; Angizi, S.; Kumar, S.; Pandey, C.M.; Simchi, A.; Willander, M.; Malhotra, B.D. Review—Textile Based Chemical and Physical Sensors for Healthcare Monitoring. *J. Electrochem. Soc.* **2020**, *167*, 037546. [CrossRef]
6. Ryu, H.; Park, S.; Park, J.J.; Bae, J. A knitted glove sensing system with compression strain for finger movements. *SMS* **2018**, *27*, 055016. [CrossRef]
7. Lee, S.; Choi, Y.; Sung, M.; Bae, J.; Choi, Y. A Knitted Sensing Glove for Human Hand Postures Pattern Recognition. *Sensors* **2021**, *21*, 1364. [CrossRef] [PubMed]
8. Isaia, C.; McMaster, S.A.; McNally, D. Study of Performance of Knitted Conductive Sleeves as Wearable Textile Strain Sensors for Joint Motion Tracking. In Proceedings of the 2020 42nd Annual International Conference of the IEEE Engineering in Medicine & Biology Society (EMBC), Montreal, QC, Canada, 20–24 July 2020; pp. 4555–4558.
9. Jansen, K.M. Performance Evaluation of Knitted and Stitched Textile Strain Sensors. *Sensors* **2020**, *20*, 7236. [CrossRef] [PubMed]
10. Bozali, B.; Ghodrat, S.; Plaude, L.; van Dam, J.J.; Jansen, K.M. Development of Low Hysteresis, Linear Weft-Knitted Strain Sensors for Smart Textile Applications. *Sensors* **2022**, *22*, 7688. [CrossRef] [PubMed]

Proceeding Paper

Smart Sock Feasibility Study †

Lucie Hernandez

Touch Craft Ltd., Penryn TR10 8NT, UK; luciefhernandez@gmail.com

† Presented at the 4th International Conference on the Challenges, Opportunities, Innovations and Applications in Electronic Textiles, Nottingham, UK, 8–10 November 2022.

Abstract: Touch Craft Ltd completed a feasibility study to understand the market position, technologies and user requirement for a smart sock sensing system designed to help monitor foot problems for people living with diabetes. The proposed system would involve a non-invasive, low-cost approach to gathering and measuring skin and foot data considering parameters such as pressure, temperature, and activity levels. The smart sock sensing system would enable both clinicians and patients to have a broader view of foot health and assist in improving the quality of life for patients and their carers.

Keywords: wearable technology; sensing system; diabetic foot; healthcare; biofeedback

1. Introduction

This paper describes a feasibility study undertaken by Touch Craft Ltd to develop a smart sock sensing system, which was supported by the Challenge Fund awarded by eHealth Productivity and Innovation in Cornwall and the Isles of Scilly [1]. The study was established to understand three key areas: the market position, technologies and user requirements for a smart sock sensing system that would be designed to help monitor foot problems.

The perspectives of clinical staff were critical in defining those people living with neuro-diabetes and foot problems as the patient population that would most benefit from a smart sock sensing system. The proposed system would help monitor foot problems in those people at high risk to help prevent ulcers and skin lesions as well as providing a foot management system to prevent re-ulceration. Healthcare professionals acknowledge the benefits of reducing foot pressures as a mechanism by which foot ulceration can be prevented [2]. Monitoring the temperature of the foot is an evidence-based preventive practice for patients at risk of diabetic foot complications [3]. The functionality of a smart sock would need to involve a non-invasive, low-cost approach to gathering and measuring skin and foot information involving pressure and temperature data.

Citation: Hernandez, L. Smart Sock Feasibility Study. *Eng. Proc.* **2023**, *30*, 11. https://doi.org/10.3390/engproc2023030011

Academic Editors: Steve Beeby, Kai Yang, Russel Torah and Theodore Hughes-Riley

Published: 29 January 2023

2. Materials and Methods

The feasibility study adopted Public and Patient Involvement and Engagement as a method of engaging patients and public in the research, referencing values and principles that should be followed to reflect good practice and guide project development [4]. Additionally, interviews with medical specialists and healthcare professionals addressed requirements for a smart sock from a clinical perspective. There was agreement from clinicians around the type of data to capture, adherence and compliance issues and the identification of monitoring and rehabilitation as significant applications for assessing and treating patients.

The researcher used PPIE methods including questionnaires, interviews and group discussion to collect data from the Lay Panel at Barts Hospital in London. The group consists of 25 members made up of patients, nurses, voluntary representatives, and clinicians that meet quarterly to improve research for the benefit of people with diabetes, giving valuable on-line feedback to researchers and refining study design and delivery [5].

2.1. Preliminary Prototypes

Touch Craft Ltd worked with Marina Toeters from By-Wire.Net [6] to design and produce innovative textile prototypes as an initial proof of concept that demonstrated core functionality (see Figure 1). They were designed to collect, process and display pressure and temperature body data using off-the-shelf technologies to illustrate the possibilities of a wearable 'smart sock' to physiotherapists and orthopaedic clinicians.

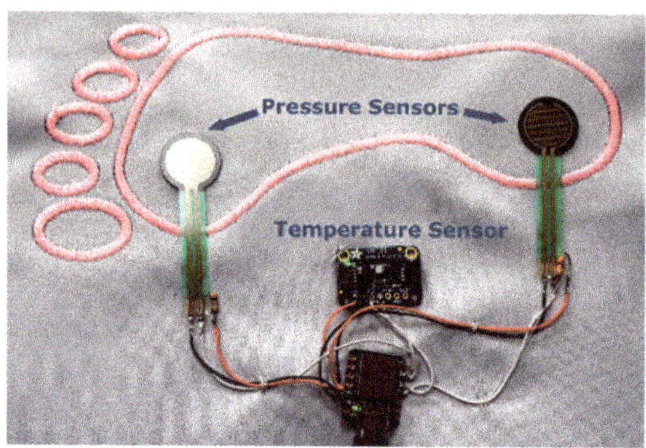

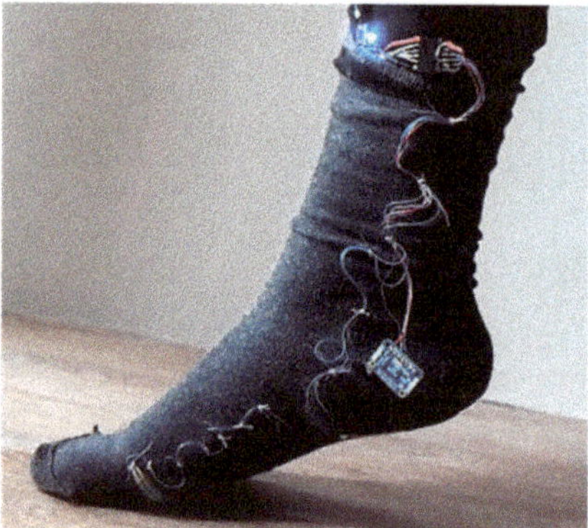

Figure 1. Preliminary prototypes.

Further funding would support the development of more refined, robust prototypes if the project were supported in future studies. At this stage, the preliminary prototypes are presented as rough sketches using 'lo-fidelity' technology to prompt discussion and help outline more specific features and functionality. A feasibility study was not extensive enough to test all the different fabrics and sensors that could be used to construct a smart sock. Prototyping helped the team understand the possibilities, limitations and outcomes that could then be experienced by clinicians and patient groups.

2.2. Printed Electronics

Printed electronics are part of an emerging market for flexible and printed sensors and components that are flexible and offer seamless integration. Printed electronics are usually screen-printed in multiple layers and fused onto fabric substrates. Touch Craft Ltd worked with Marina Toeters from the fashion and technology company By-Wire.Net [6] to demonstrate the viability of printed electronic components that could be included in future prototypes (see Figure 2).

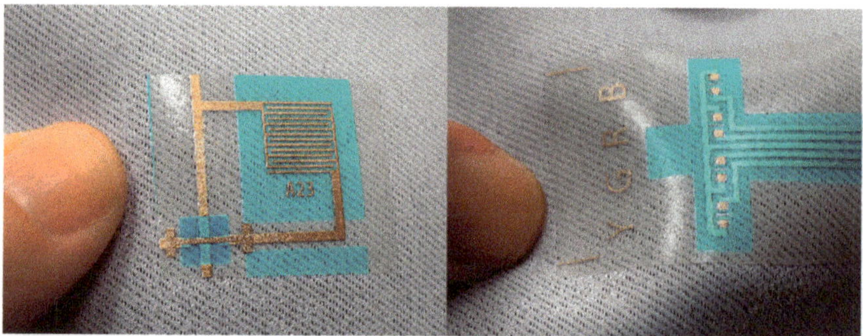

Figure 2. Printed electronics.

3. Results

A summary of themes that emerged from interviews and questionnaires are presented below and have been organised into sections depending on their overall theme.

3.1. Remote Consultations

- Patient groups wanted to maintain in-person consultations with clinical teams as a move against the increased interest in remote sensing and remote consultations.
- Patient groups were concerned that the increased prevalence of remote consultations would replace the foot health checks and examinations conducted by professionals using a variety of methods; e.g., filament, ultrasound, colour checks and circulation.

3.2. Foot Health, Mobility and Compliance

- The study highlighted a general lack of concern for foot health, not just in diabetic patients, and the technical difficulties involved in older adults monitoring their own foot issues.
- Clinicians were keen to stress the importance of patients taking a more active role to identify and deal with problems relating to their own foot health.
- Mobility is a crucial part of wellbeing and overall good health, and there was some concern that sensing technology embedded within a smart sock would not be able to detect mobility issues and alert patients.
- There was optimism among clinical professionals that the device (particularly temperature sensing) could have practical applications for remote monitoring and self-management of diabetic feet at risk of re-ulceration.

3.3. Practical Considerations, Design, Ergonomics and Fit

- Patient groups wanted a smart sock to be comfortable to wear and a good fit, preferably seamless, so that seams would not chafe on the toes.
- Patient groups requested more details on washing cycles and the standard length of time advised for smart sock use.
- Respondents questioned if the device would be suitable for wearing inside a shoe.

3.4. System Design, Technical Features and Data Transmission

- Many respondents wanted to know if a smart sock would be able to detect excesses of foot heat, cold, numbness, sweat, smell and fragrance.
- Patient groups were concerned with issues of intermittent signal or signal loss during data transmission or communication with clinicians.
- Respondents were keen to know more about the kind of feedback provided by the system to alert patients or clinicians, perhaps a 'bleep', an alarm or colour change on the sensor to indicate loss of feeling in the feet.
- Clinicians reported on the position and number of the pressure sensors on the foot. For more accurate pressure sensing, the sensors would need to be embedded across the whole forefoot.

3.5. Funding, Development and Future Studies

- Respondents raised concerns around the project being a beneficial investment and requested more information about methods of funding smart sock development.

4. Discussion

Many of the findings outlined themes that could be explored in future studies but were outside the scope of this limited feasibility study. Patient and clinical perspectives emphasized technical as well as practical elements, which emerged as strong themes and demonstrated a concern with the design of the sock to maximise comfort, fit and performance. More details were requested regarding the technical features and how these could be more seamlessly embedded into the sock. Much of the future work in this area would concentrate on integrating printed electronics that are discussed in Section 3.2.

What emerged from the feedback was that many of the respondents were interested in the idea of collecting foot information and biofeedback to indicate issues around foot health. Patients were particularly interested in the idea of exploring further the idea of developing a 'digital health assistant' that could support patients in managing their own foot care.

5. Conclusions

Patient feedback indicated that a future smart sock should not replace in-person consultations but would complement the personal, reassuring meetings with health professionals. This would be enhanced with data gathered from home monitoring and self-management of foot problems. This might include conducting virtual foot checks as well as regular manual foot checks to monitor changes and risk factors.

To conclude, it is worth noting that more robust, hi-fidelity prototypes would not have been produced at this stage in the development cycle. The feasibility study aimed to present information to support the proposition for the smart sock concept being technically and clinically viable and identifying patient support for the idea. The study was successful in determining a clinical and patient need for a smart sock sensing system and suggesting that further funding and research could progress the idea into a robust and beneficial product.

Funding: This project received a feasibility grant awarded by EPIC, eHealth Productivity and Innovation in Cornwall and the Isles of Scilly, funded by the European Regional Development Fund.

Institutional Review Board Statement: Ethical review and approval were waived for this study due to the restricted nature of the research methods adopted.

Informed Consent Statement: Patient consent was waived as it was categorised as involvement rather than research because it asked opinion to inform what was proposed rather than collecting study data.

Data Availability Statement: The data presented in this study are available on request from the corresponding author. The data are not publicly available due to the conditions of the funder.

Acknowledgments: The author would like to thank Joanne Paton, School of Health Professions, Faculty of Health: Medicine, Dentistry and Human Sciences, University of Plymouth.

Conflicts of Interest: The author declares no conflict of interest.

References

1. eHealth Productivity and Innovation in Cornwall and the Isles of Scilly (EPIC). Available online: https://www.plymouth.ac.uk/research/epic (accessed on 1 December 2022).
2. Collings, R.; Freeman, J.; Latour, J.M.; Paton, J. Footwear and insole design features for offloading the diabetic at risk foot—A systematic review and meta-analyses. *Endocrinol. Diabetes Metab.* **2020**, 4, e00132. [CrossRef] [PubMed]
3. Lavery, L.A.; Petersen, B.J.; Linders, D.R.; Bloom, J.D.; Rothenberg, G.M.; Armstrong, D.G. Unilateral remote temperature monitoring to predict future ulceration for the diabetic foot in remission. *BMJ Open Diab. Res. Care* **2019**, 7, e000696. [CrossRef] [PubMed]
4. INVOLVE. *Values, Principles and Standards for Public Involvement in Research*; INVOLVE: Eastleigh, UK, 2015.
5. Diabetes Research Lay Panel. Available online: https://www.qmul.ac.uk/blizard/research/research-groups/barts-diabetes-obesity-research-group/diabetes-research-lay-panel/ (accessed on 2 December 2022).
6. By-Wire.Net. Available online: https://www.by-wire.net/ (accessed on 28 November 2022).

Proceeding Paper

Effect of Bandage Materials on Epidermal Antenna †

Irfan Ullah [1,*], Mahmoud Wagih [2], Abiodun Komolafe [1] and Steve P. Beeby [1]

1 School of Electronics and Computer Science, University of Southampton, Southampton SO17 1BJ, UK
2 James Watt School of Engineering, University of Glasgow, Glasgow G12 8QQ, UK
* Correspondence: i.ullah@soton.ac.uk

† Presented at the 4th International Conference on the Challenges, Opportunities, Innovations and Applications in Electronic Textiles, Nottingham, UK, 8–10 November 2022.

Abstract: This study explores the effect of different types of bandages on the performance of an epidermal antenna. Three identical dipole antennas are designed on three different types of bandages, and the measured reflection coefficients, S_{11}, show that the antennas resonate at the same frequency despite the different types of fabric bandages. However, the antennas resonance frequency shifts to a lower frequency when the antennas are mounted on the body. The transmission coefficient, S_{21}, over a 60 cm link with a standard RFID antenna is at least −30 dB, and −34 dB in free space and on the body, respectively, demonstrating that the antenna is suitable for communication and wireless RF power transfer in wearable applications.

Keywords: epidermal antenna; dipole; smart bandage; e-textiles; wearable application

1. Introduction

Smart bandages incorporate various types of sensors to continuously monitor of wound−related parameters, such as temperature, moisture level, pH level, and wound oxygenation, in chronic wound care and management [1,2]. They provide wound data to health practitioners, which allows them to remotely assess the healing of chronic wounds without removing the bandage. The smart bandage requires a power source for embedded electronics, and an antenna for wireless data transmission to an external device. An antenna design is critical in the development of a wireless smart bandage since it can be used to transmit data and harvest RF energy.

Several antenna designs for smart bandages have been presented in the literature. In [3], a via free planar antenna, similar to an adhesive bandage, for medical telemetry service is proposed. However, the antenna includes a ground plane, which increases the thickness of the antenna and, therefore, is less suitable for wearable applications. Similarly, in [4], a planar rectangular loop antenna is implemented in a battery−powered smart bandage for wireless monitoring of wounds. The antenna is small in size but operates at a higher frequency of about 2.4 GHz. Furthermore, near−field communications (NFC) antennas are also being investigated for wireless smart bandages [5]. Such bandages have a very low reading range, and need the bandage to be in close proximity to the reader, which is especially undesirable for applications requiring continuous monitoring.

In this study, we proposed an all−fabric epidermal antenna operating at 915 MHz for smart bandages in healthcare applications. The study also explored the effect of different types of bandages on the epidermal antenna resonance frequency. Three identical dipoles were designed on three different types of bandage materials. The performance of the antennas in terms of the reflection coefficient in free space and in the presence of body were investigated. Measurements demonstrate that the different types of bandage material have no discernible effect on the antenna performance.

Citation: Ullah, I.; Wagih, M.; Komolafe, A.; Beeby, S.P. Effect of Bandage Materials on Epidermal Antenna. *Eng. Proc.* **2023**, *30*, 12. https://doi.org/10.3390/engproc2023030012

Academic Editors: Steve Beeby, Kai Yang, Russel Torah and Theodore Hughes-Riley

Published: 29 January 2023

2. Antenna Design

The epidermal antenna design is based on an electrical dipole with bending radiating arms to reduce the antenna size. The radiating arms of the antenna were tuned to resonate at 915 MHz in the presence of human tissue. The antennas were made of silk coated Litz wires with a diameter of 0.36 mm. A PFAFF creative 3.0 sewing machine was used to embroider Litz wires into three types of bandages: (i) cotton crepe bandage made of cotton, (ii) self−fixing cohesive support bandage made of cotton/elastane with latex, and (iii) adjustable cohesive bandage made of polypropylene and elastane, as shown in Figure 1.

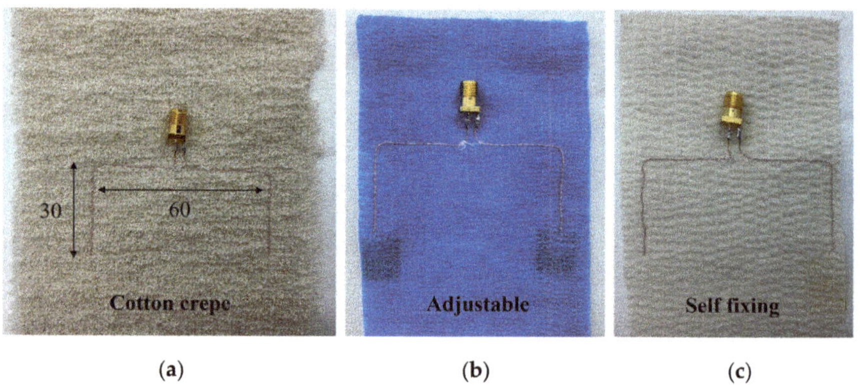

Figure 1. Manufactured prototype of the embroidered all−fabric epidermal antennas: (**a**) cotton crepe bandage; (**b**) adjustable bandage; and (**c**) self−fixing bandage.

3. Measurements and Results

3.1. Reflection Coefficient

The reflection coefficient, S_{11}, of the antennas were measured with a vector network analyser (VNA) in free space and in the presence of the body, Figure 2a. Figure 2b shows that the antennas resonate at around 1.10 GHz in free space, and 915 MHz when the antennas are mounted on the body. It is observed that the different types of bandages have no significant effect on the antenna resonance frequency, as shown in Figure 2b. However, the resonance frequency shifts to a lower frequency of 915 MHz in the presence of the body. This is due to the high dielectric constant and conductivity of human tissue.

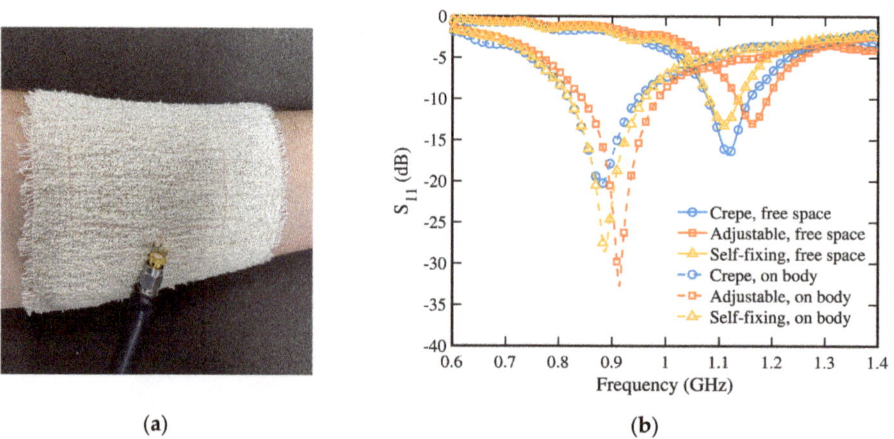

Figure 2. (**a**) All−fabric epidermal antenna mounted on the human arm; (**b**) measured reflection coefficient, S_{11}, of the three identical antennas in free space and on the body.

3.2. Transmission Coefficient

The experimental setup, Figure 3a, was used to measure the transmission coefficient, S_{21}, between the fabric antenna and an external antenna. A circularly polarized antenna, with 8.2 dBi gain and operating at 915 MHz, was placed about 60 cm away from the bandage antenna. Both antennas were connected to a vector network analyser (VNA) and the transmission losses were measured. The measured S_{21} frequency responses are depicted in Figure 3b for both cases with and without human tissue. The results show that S_{21} is about −30 dB in free space, and −34 dB when mounted on the body. This shows that for a 30 dBm RF input power, at least −4 dBm will be received by the receiver antenna, indicating that the bandage antenna is suitable for RF power harvesting over a short distance at the UHF band.

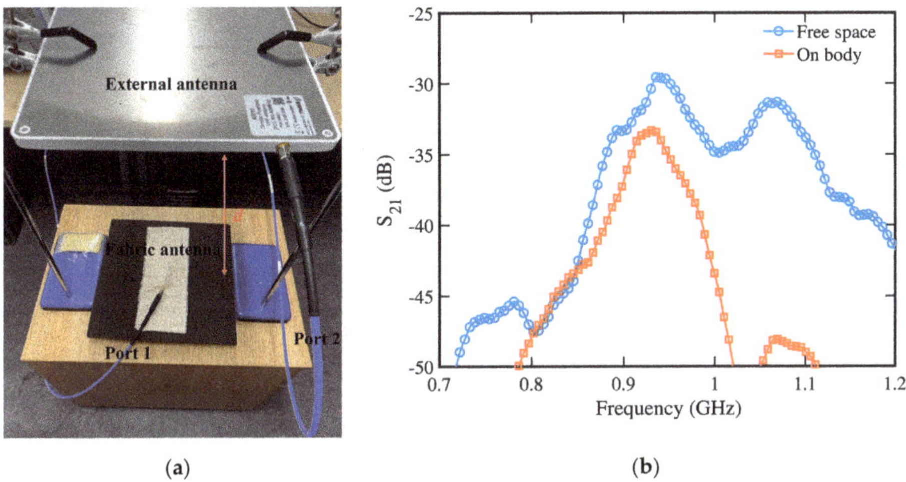

(**a**) (**b**)

Figure 3. (**a**) Experimental setup for measuring transmission performance; (**b**) measured transmission coefficient, S_{21}, between the all−fabric antenna and an external antenna, showing that the transmission losses for the antenna in free space and on the body are −30 dB and −34 dB, respectively.

4. Conclusions

All-fabric epidermal antenna fabricated on fabric bandages is demonstrated in this paper. The resonance frequency of the antenna shifts to a lower frequency when it is mounted on the body due to the high relative permittivity and conductivity of human tissue. The measured results show that the different types of bandages have no significant effect on the antenna resonance frequency. Transmission losses, S_{21}, of the antenna is −34 dB in the presence of human tissue when the external antenna is 60 cm away from the arm on which the antenna is mounted. This means that the receiver antenna will receive at least −4 dBm for an input power of 30 dBm. The antenna is flexible, lightweight, easy to fabricate, and comfortable to the body, and, therefore, can be used to develop wireless and battery−free smart bandages. The objective of our future work is to closely investigate all fabric dipole array for RF information and RF power transfer for wearable applications.

Author Contributions: Conceptualization, I.U. and M.W.; methodology, I.U.; software, I.U.; validation, I.U., M.W., A.K., and S.P.B.; formal analysis, I.U.; investigation, I.U.; writing—original draft preparation, I.U.; writing—review and editing, M.W. and A.K.; supervision, S.P.B.; project administration, S.P.B.; funding acquisition, S.P.B. All authors have read and agreed to the published version of the manuscript.

Funding: This research was funded by European Regional Development Fund (ERDF) via its Inter-reg V France (Channel) England programme: Smart Textile for Regional Industry and Smart Specialisation Sectors (SmartT). The work of Steve Beeby was supported by the Royal Academy of Engineering under the Chairs in Emerging Technologies Scheme.

Institutional Review Board Statement: Not applicable.

Informed Consent Statement: Not applicable.

Data Availability Statement: Not applicable.

Acknowledgments: We thank Yixuan Sun for helping in the fabrication of antennas on the bandages.

Conflicts of Interest: The authors declare no conflict of interest.

References

1. Mostafalu, P.; Lenk, W.; Dokmeci, M.R.; Ziaie, B.; Khademhosseini, A.; Sonkusale, S.R. Wireless flexible smart bandage for continuous monitoring of wound oxygenation. *IEEE Trans. Biomed. Circuits Syst.* **2015**, *9*, 670–677. [CrossRef] [PubMed]
2. Escobedo, P.; Bhattacharjee, M.; Nikbakhtnasrabadi, F.; Dahiya, R. Smart bandage with wireless strain and temperature sensors and batteryless NFC tag. *IEEE Internet Things J.* **2021**, *8*, 5093–5100. [CrossRef]
3. Chi, Y.; Chen, F. On-body adhesive-bandage-like antenna for wireless medicaltelemetry service. *IEEE Trans. Antennas Propag.* **2014**, *62*, 2472–2480. [CrossRef]
4. Farooqui, M.; Shamim, A. Low cost inkjet printed smart bandage for wireless monitoring of chronic wounds. *Sci. Rep.* **2016**, *6*, 28949. [CrossRef] [PubMed]
5. Li, Y.; Grabham, N.; Komolafe, A.; Tudor, J. Battery free smart bandage based on NFC RFID technology. In Proceedings of the IEEE International Conference on Flexible and Printable Sensors and Systems (FLEPS), Manchester, UK, 16–19 August 2020; pp. 1–4.

Proceeding Paper

Novel Strain Sensor in Weft-Knitted Textile for Triggering of Functional Electrical Stimulation †

Bahareh Abtahi [1,2,*], Mareen Warncke [1,2], Hans Winger [1,2], Carmen Sachse [1], Eric Häntzsche [1], Andreas Nocke [1,2] and Chokri Cherif [1,2]

1 Institute for Textile Machinery and High Performance Material Technology (ITM), Faculty for Mechanical Science and Engineering, Technische Universität Dresden, 01069 Dresden, Germany
2 CeTI—Cluster of Excellence, Centre for Tactile Internet with Human-in-the-Loop, Technische Universität Dresden, 01069 Dresden, Germany
* Correspondence: bahareh.abtahi@tu-dresden.de; Tel.: +49-(0)-351-463-31915
† Presented at the 4th International Conference on the Challenges, Opportunities, Innovations and Applications in Electronic Textiles, Nottingham, UK, 8–10 November 2022.

Abstract: Functional electrical stimulation (FES) aims to improve the gait pattern in case of foot drop of people suffering chronic diseases, e.g., multiple sclerosis. The fibular nerve can be stimulated by electrical impulses sent through electrodes on the skin, which leads to the contraction of the corresponding muscles. One major disadvantage of commercial FES devices is their bulky design. The paper presents an alternative approach of weft-knitted strain sensors that are directly integrated into the knee area of a functional legging suitable for daily use. To initiate electrical impulses for FES at the right time, the textile strain sensors are used as soft triggers.

Keywords: weft-knitted fabric; textile-based strain sensor; functional electrical stimulation; multiple sclerosis; functional electrical stimulation; FES

Citation: Abtahi, B.; Warncke, M.; Winger, H.; Sachse, C.; Häntzsche, E.; Nocke, A.; Cherif, C. Novel Strain Sensor in Weft-Knitted Textile for Triggering of Functional Electrical Stimulation. *Eng. Proc.* **2023**, *30*, 13. https://doi.org/10.3390/engproc2023030013

Academic Editors: Steve Beeby, Kai Yang, Russel Torah and Theodore Hughes-Riley

Published: 31 January 2023

1. Introduction

Wearables have developed very rapidly over the last decade and have become integral parts of our daily lives. Due to their convenience and flexibility, they are used for gesture and posture tracking in healthcare and biomechanical and physiological monitoring systems [1]. An important functional therapy method for multiple sclerosis (MS) related foot drop or stroke patients is functional electrical stimulation (FES), in which the muscle is functionally activated by electrical nerve stimulation and can thus perform its physiological movement [2]. Neuro-prostheses are one of the advances that have been made in the field of FES therapy. They detect the gait phases through inertial measurement units (IMU), which then trigger electrical impulses to stimulate the corresponding nerve at the appropriate moment so that the foot rises [3,4]. Commercially available assistance systems for people with foot-lift weakness are, e.g., L100/300 Go® [5,6] from Bioness, innoSTEP-WL [7] from Heller Medizintechnik, or WalkAide® from Pro Walk [8]. In these bulky devices, the sensors and electrical signal generator are integrated directly into a cuff that is fixed with Velcro straps around the knee. However, the size and mass, the built-in hard electronics (inflexible, non-bendable areas), and the limited washability are comfort-limiting factors of these devices for daily use. Therefore, the aim of the current research at ITM is the realization of a textile-based sensory and therapeutic stimulative functional system integrated into a weft-knitted fabric in the form of functional leggings.

Electrically conductive yarns are suitable for the realization of resistive strain sensors, whose electrical resistance changes in correlation to the applied mechanical strain and can be integrated into a weft-knitted fabric during manufacturing. In addition to that, the joint movement according to the natural degrees of freedom that causes a corresponding strain state in the textile can also be measured. Due to the loop structure, knitted fabrics are stretchable in both horizontal and vertical directions, resulting in a low bending modulus and shear. The user comfort of weft-knitted fabrics in terms of freedom of movement make them a suitable choice for use as supporting structures of wearable healthcare products [9], e.g., for both future textile-based strain sensors and dry electrodes for FES therapy purposes, respectively. Due to the loop structure, weft-knitted fabrics are also characterized by excellent permeability and flexibility. The gait analysis for FES devices requires a real-time diagnostic function to determine the movements based on the deformation measurement, for which integrated textile strain sensors are predestined. The textile-based strain sensor is positioned on the kneecap and is intended to determine valid motion-induced strain data without restricting the freedom of movement of the functional leggings. The textile structure is stretched accordingly by the bending movement of the leg during the gait phase, and the measurable change in ohmic resistance can be correlated with the movement phase. Based on previous investigations using weft-knitted structures, the maximum expected elongation in the knee area is determined to a max of 50%.

2. Materials and Methods

For the gait analyses, suitable conductive yarn-based resistive strain sensors on the knee area are needed. Based on investigations using weft-knitted structures, the maximum expected elongation in the knee area is determined to the max of 50%. It is essential that the textile sensor is flexible and does not restrict free movement. With dynamic movement, the knee or bent angle changes continuously. Bending is defined here as the direction of motion, which leads to the reduction of angle and stretching, causing an increase of the angle. To determine the expected length change of the weft-knitted structure (leggings) above the kneecap, the knee is bent in 45° steps from the unbent to full-bent state and the lengths L_1 to L_4 are measured. Based on the observations, the expected elongation of weft-knitted structures over the knee joint during walking is determined at a range from 23 to 50%. This represents the measuring range of the textile strain sensor (cf. Table 1). Although other gait analyses come to similar conclusions that the bending/flexion of the knee joint during normal walking reaches max. 60° (max. elongation/extension 0°), for the further development steps [10,11], a maximum elongation of 30% is assumed as the average deformation to be expected during the normal gait phase.

Table 1. Determined elongation of a weft-knitted structure fixed over knee joint during leg angulation.

Leg Position, Bent Angle	Length Knit Structure L_i [cm]		Expected Strain $\varepsilon_m = L_i\text{-}L_{i\text{-}1}/L_{i\text{-}1}$ [%]
Basic length	L_0	9.2	-
Pre-stretched, unbent 0°	L_1	12	23
Bent by 45°	L_2	15.1	26
Bent by 90°	L_3	16.3	36
Full bent (140°)	L_4	18	50

Weft-knitted strain sensors are realized on a conventional weft-knitting machine (Karl Mayer Stoll ADF 530-32 BW knit & wear, E14 machine gauge) available at ITM. The basic knit structures have been developed, and material combinations were evaluated. The investigations were based on a plated, plain, right—left weft-knitted structure (single jersey) made of a material combination from TENCEL™ lyocell fibers (fineness 25 tex) plated with a polyamide and elastane twine PA/EL 78/78f23x1. The strain sensors are integrated into four different bindings in wale-wise (A, B) and course-wise directions (C, B). The strain-sensor pattern was realized by intarsia knitting or platting, respectively, using different silver coated/plated conductive yarns Silver-tech+ 150, Elitex 235/f36 PA/Ag. Table 2 shows the investigated materials and bindings of the weft-knitted basic structure and the textile-based sensors.

Table 2. Overview materials and bindings of basic weft-knitted structure and textile sensors.

Sensor Yarns (Red Area)	Binding Basic Knit and Material (Yellow Area)	Sensor Binding	
I: Silver-tech+ (Silver coated pol- yamide yarn; 150 tex) II: Elitex 235/f36 (Silver coated pol- yamide yarn)	Plated right left single jersey Tencel 25 tex plated with PA/EL single cover yarn 78/78f23x1	A B C D	Wale-wise plating of sensor yarns A: 4 wales × 200 courses B: 1 wale × 200 courses course-wise plating of sensor yarns C: 100 wales × 4 courses D: 100 wales × 1 course

For characterization of the strain sensor's cyclical tensile behavior, specimens (cf. Table 2) were tested on a ZwickRoell Junior Z2.5 tensile tester. The tensile strain is increased from 0 to 50% successive in 10% steps after 10 cycles while measuring the change in resistance of the textile-based strain sensors simultaneously with a Keithley DAQ 6510 precision multimeter in a 4-wire method.

3. Results and Discussion

As observed in Figure 1a, the sensor binding A with yarn I has a good correlation between change in strain and resistance compared with sensor C (cf. Figure 1b) with the same yarn (cf. Figure 1b). The hysteresis increases for strain levels above 25% significantly, as Figure 1c shows, and the cross-correlation coefficient was calculated to 0.83. Furthermore, a latent signal drift after each load cycle within a constant elongation level can be observed.

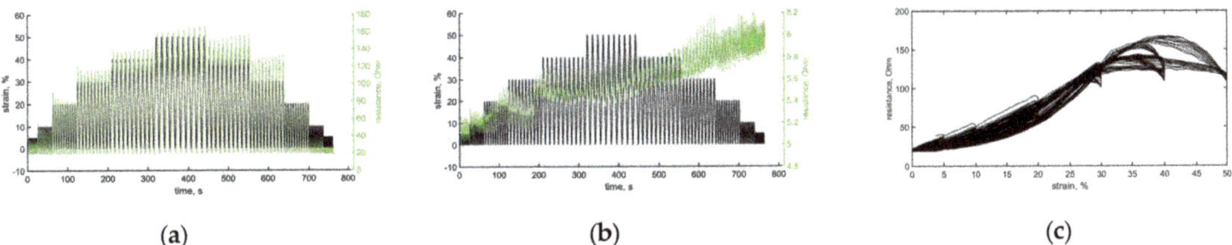

(a) (b) (c)

Figure 1. Electromechanical behavior of sensor yarn I during cyclic straining: (**a**) change in resistance with binding A; (**b**) change of resistance with binding C; (**c**) resistance-strain correlation for binding A.

Figure 2 shows that sensor binding A with yarn II also has a good correlation between the change in resistance and applied strain with a cross-correlation coefficient 0.94 (cf. Figure 1c). For sensor binding C (cf. Figure 2b), which means the course-wise integration of the sensor yarn in a single wale over 100 courses, the resistance-strain correlation is remarkably low.

Figure 3 shows that the sensor binding B with yarn I (cf. Figure 3a,c) has the best correlation between strain and resistance. The cross-correlation coefficient was calculated to 0.99. For all investigated sensor yarns, sensor bindings C and D (cf. Figure 3b) show the lowest resistance-strain correlation of the investigated test series. It can be stated that the course-wise platting of conductive yarns is not suitable for reliable strain sensing.

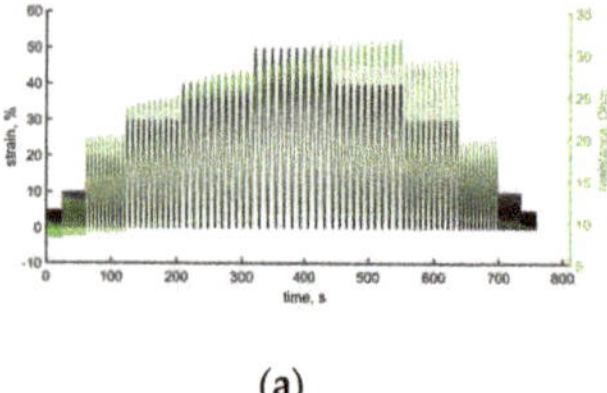

(a)

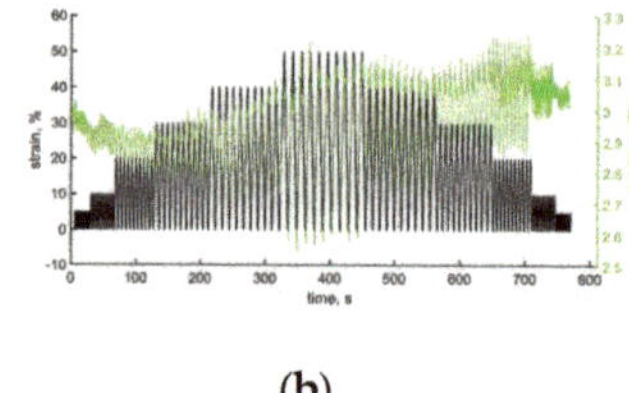

(b)

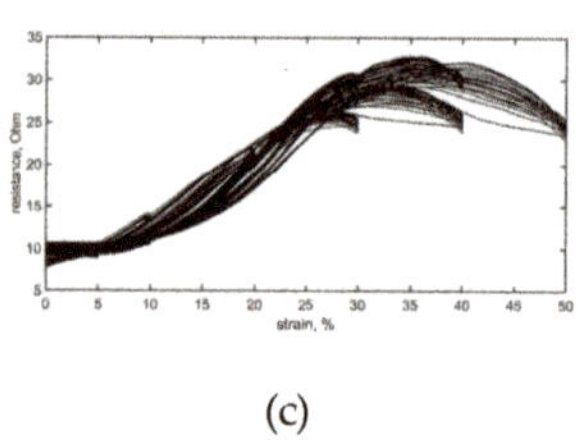

(c)

Figure 2. Electromechanical behavior of sensor yarn II during cyclic straining: (**a**) change in resistance with binding A; (**b**) change of resistance with binding C; (**c**) resistance−strain correlation for binding A.

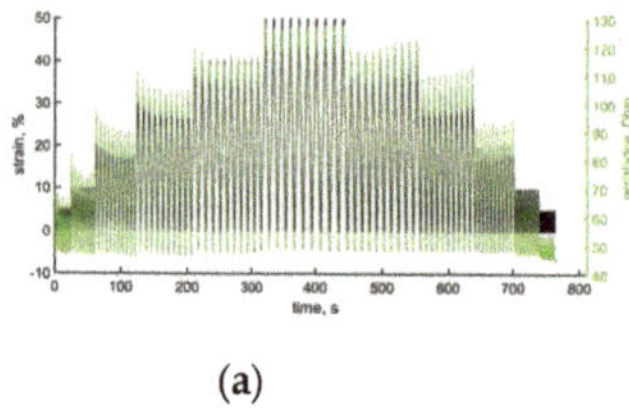

(a)

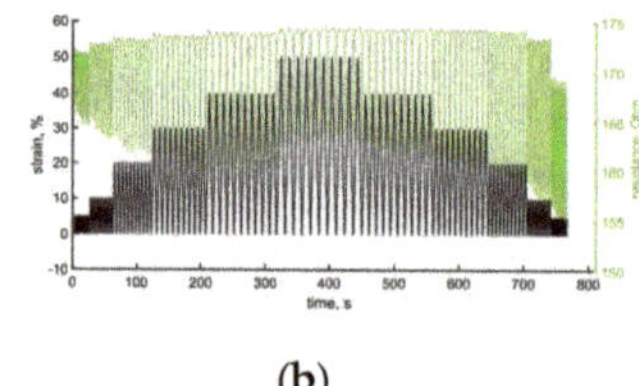

(b)

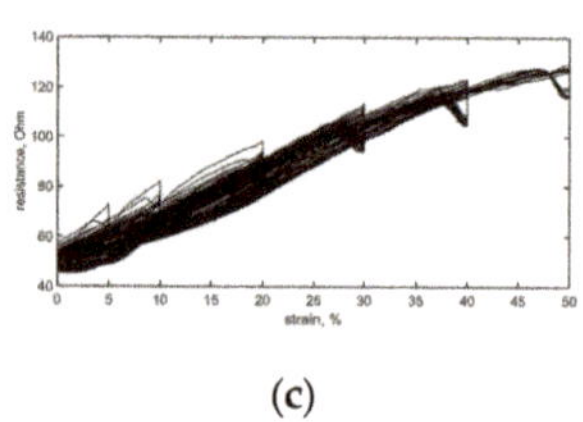

(c)

Figure 3. Electromechanical behavior of sensor yarn I during cyclic straining: (**a**) change in resistance with binding B; (**b**) change of resistance with binding D. (**c**) resistance−strain correlation for binding B.

4. Conclusions

This work presents a novel approach for the integration of conductive yarns into weft-knitted fabric acting as strain sensor for monitoring and detecting the gait phase for FES applications, e.g., for patients suffering from MS. Two conductive yarns in wale-wise and course-wise directions have been integrated into a plated plain right-left weft-knitted structure, and the electromechanical behavior was determined during cyclical tensile tests. It can be concluded that the strain sensor made of silver-coated yarn Silver-tech+ 150 plated with PA/EL elastane twine in a plain right/left wale-wise binding structure B (1 wale × 200 courses) shows the best result concerning a change of resistance under an applied cyclic tensile strain up to 30% with cross-correlation coefficient calculated to 0.99.

Author Contributions: Conceptualization, B.A., H.W., and E.H.; methodology, B.A. and M.W.; software, M.W.; knitting design, knitting, specimen preparation, C.S.; validation, E.H., A.N., and M.W.; formal analysis, M.W.; investigation, B.A. and H.W.; resources, B.A.; data curation, M.W.; writing—original draft preparation, B.A.; writing—review and editing, E.H.; visualization, B.A.; supervision, E.H.; project administration, C.C.; funding acquisition, E.H. All authors have read and agreed to the published version of the manuscript.

Funding: The project is funded by the German Research Foundation (DFG. Deutsche Forschungsge meinschaft) as part of Germany's Excellence Strategy—EXC 2050/1—funding Nr. 390696704—Cluster of Excellence "Centre for Tactile Internet with Human-in-the-Loop" (CeTI) of Technische Universität Dresden. The ZIM project "SmartMediTex", funding Nr. KK5090906SK1, is funded through the AiF within the "Central Innovation Program for small and medium-sized enterprises (SMEs)—Zentrales Innovationsprogramm Mittelstand (ZIM)—Cooperation projects" from funds of the Federal Ministry for Economic Affairs and Climatic Action (BMWK) by a resolution of the German Bundestag. Funding of the named institutions is gratefully acknowledged.

Institutional Review Board Statement: Not applicable.

Informed Consent Statement: Not applicable.

Data Availability Statement: All data are available from authors by request.

Conflicts of Interest: The authors declare no conflict of interest.

References

1. Isaia, C.; Mcnally, D.S.; Mcmaster, S.A.; Branson, D.T. Effect of mechanical preconditioning on the electrical properties of knitted conductive textiles during cyclic loading. *Text. Res. J.* **2019**, *89*, 445–460. [CrossRef]
2. Miller, L.; Mattison, P.; Paul, L.; Wood, L. The effects of transcutaneous electrical nerve stimulation (TENS) on spasticity in multiple sclerosis. *Mult. Scler.* **2007**, *13*, 527–533. [CrossRef] [PubMed]
3. Gorman, P.H. An Update on Functional Electrical Stimulation after Spinal Cord Injury. *Neurorehabilit. Neural Repair* **2000**, *14*, 251–263. [CrossRef] [PubMed]
4. Creasey, G.H.; Ho, C.H.; Triolo, R.J.; Gater, D.R.; DiMarco, A.F.; Bogie, K.M.; Keith, M.W. Clinical applications of electrical stimulation after spinal cord injury. *J. Spinal Cord Med.* **2004**, *27*, 365–375. [CrossRef] [PubMed]
5. Hausdorff, J.M.; Ring, H. The effect of the L300 neuroprosthesis on gait stability and symmetry. *J. Neurol. Phys. Ther.* **2006**, *30*, 198–199. [CrossRef]
6. Improved Mobility. Made Easier. Bioness Inc., NC 27703, United States: 2022. Available online: https://www.l300go.com/ (accessed on 21 November 2022).
7. HELLER MEDIZINTECHNIK GmbH & Co. KG: Weak Foot Dorsi Flexion—Foot Drop System innoSTEP-WL Provides Mobility. Available online: https://www.heller-medizintechnik.de/produkte/innostep_wl/?lang=EN (accessed on 12 November 2022).
8. Pro Walk Rehabilitationshilfen und Sanitätsbedarf GmbH: The WalkAide®—System. Myo-Orthetic Technology for the Treatment of Centrally Caused Foot Lift Weakness. Available online: https://www.prowalk.de/produkte/walkaide/ (accessed on 20 November 2022).
9. Euler, L.; Guo, L.; Persson, N.-K. Textile Electrodes: Influence of Knitting Construction and Pressure on the Contact Impedance. *Sensors* **2021**, *21*, 1578. [CrossRef] [PubMed]
10. Watson, A.; Sun, M.; Pendyal, S.; Zhou, G. TracKnee: Knee angle measurement using stretchable conductive fabric sensors. *Smart Health* **2020**, *15*, S.100092f. [CrossRef]
11. Götz-Neumann, K. *Understanding Walking—Gait Analysis in Physiotherapy*, 4th ed.; Thieme; Georg Thieme Verlag: Stuttgart, Germany, 2016; ISBN 9783132401549.

engineering proceedings

Proceeding Paper

Developing High-Resolution Thin-Film Microcircuits on Textiles †

Abiodun Komolafe [1,*], Michael Gakas [2] and Steve Beeby [1]

1 Centre of Flexible Electronics and E-Textiles, University of Southampton, Southampton SO17 IBJ, UK
2 TE Connectivity, Chem. des Chapons-des-Prés 11, 2022 Bevaix, Switzerland
* Correspondence: a.o.komolafe@soton.ac.uk
† Presented at the 4th International Conference on the Challenges, Opportunities, Innovations and Applications in Electronic Textiles, Nottingham, UK, 8–10 November 2022.

Abstract: Scaling down the form-factor of printed electronics is one of the methods for improving the reliability of printed e-textiles. This also enhances the wearability of the printed e-textile. However, the surface roughness of textiles and the low resolution of current printing methods, such as screen-printing, often present significant challenges for directly realizing microcircuits on textiles that are developed for printed e-textile applications. This work reports the planarization of a polyester cotton textile with a screen-printed polyurethane (PU) smoothing interface layer to enable the micro-patterning of the textile with conductive thin films using microfabrication techniques. Thermally evaporated copper structures with features sized from 800 μm down to 10 μm are patterned on the planar textile, demonstrating a printed resolution that is otherwise difficult to achieve through screen-printing even with the aid of specialized screens.

Keywords: e-textiles; microfabrication techniques; polyurethane interface layer; micropatterns; surface roughness; reliability

1. Introduction

The challenge of manufacturing lightweight e-textiles that are unobtrusive and robust against external stresses continues to drive the research and applications of e-textiles. Meeting this need necessitates a reduction in the geometry and size of any integrated electronics on textiles down to the micron scale (<100 μm) [1]. However, the feature resolution of current e-textile manufacturing methods such as printing [2], embroidery [3] and weaving [4] is insufficient to achieve such microcircuits on textiles. Microcircuits for e-textiles are currently manufactured using traditional microfabrication techniques such as photolithography and etching or lift-off processes to create flexible electronic strip circuits suitable for weaving into fabrics [5]. During the weaving manufacturing stresses, this method frequently introduces stresses into the microcircuit. The weaving process further complicates the integration of these strip circuits in fabrics [6].

This study offers preliminary research into an approach that combines low-cost screen-printing with standard microfabrication processes to directly deposit and localize microcircuit patterns on textiles, minimizing integration challenges and benefiting the reliability and wearability of e-textiles. This method also ensures that micropatterns are not strained by the manufacturing process. The proposed e-textile shown in Figure 1 consists of a screen-printed polyurethane (PU) interface layer, which planarizes the textile surface and allows copper micropatterns to be deposited onto it through a combination of photolithography, thermal evaporation deposition and etching processes.

Citation: Komolafe, A.; Gakas, M.; Beeby, S. Developing High-Resolution Thin-Film Microcircuits on Textiles. *Eng. Proc.* **2023**, *30*, 14. https://doi.org/10.3390/engproc2023030014

Academic Editors: Kai Yang, Russel Torah and Theodore Hughes-Riley

Published: 31 January 2023

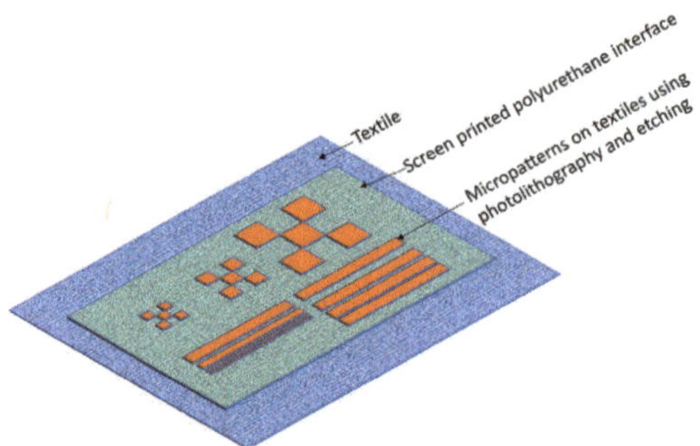

Figure 1. Proposed e-textile fabricated using screen-printing and microfabrication techniques.

2. Materials and Methods

Polyester cotton fabric supplied by Klopmann Ltd. was chosen as the textile substrate due to its suitability for garment manufacture. Screen-printable UV paste, UV-IF-1004, supplied by Smart Fab inks was used for printing the smoothing polyurethane (PU) interface layer on the fabric because of its good adhesion to printed films [1]. Table 1 lists the solvents used in the lift-off or etching microfabrication processes for patterning the textile.

Table 1. List of solvents for lift-off and etching processes.

Microfabrication Process	Chemicals/Solvents	Purpose
Etching process	AZ nLOF 2070 photoresist	Negative resist solution
	AZ 726 MIF Developer solution	To develop AZ nLOF 2070 after UV-exposure
	Ferric Chloride solution	To etch copper film
	Chrome etchant UN0398	To etch chromium film
Lift-Off process	AZ 9260 photoresist	Positive resist solution
	AZ 400K developer	To develop positive resist after UV-exposure
	N-Methyl-2-pyrrolidone (NMP)	To strip positive resist after metallization of substrate
	Acetone	To strip positive resist after metallization of substrate
General	De-ionized water	To clean substrate after developing

2.1. Screen-Printing and Metallization Processes

Figure 2 shows a 10 cm × 10 cm PU interface layer screen-printed onto the textile using a semi-automatic DEK248 screen-printer with the printing process described in [1]. The average printed PU thickness was 200 μm. The 2D surface topography of the printed textile obtained from a Tencor P11 surface profiler, as shown in Figure 2, clearly indicates a significant reduction in the average surface roughness of the textile from an initial value of 35 μm to 1 μm.

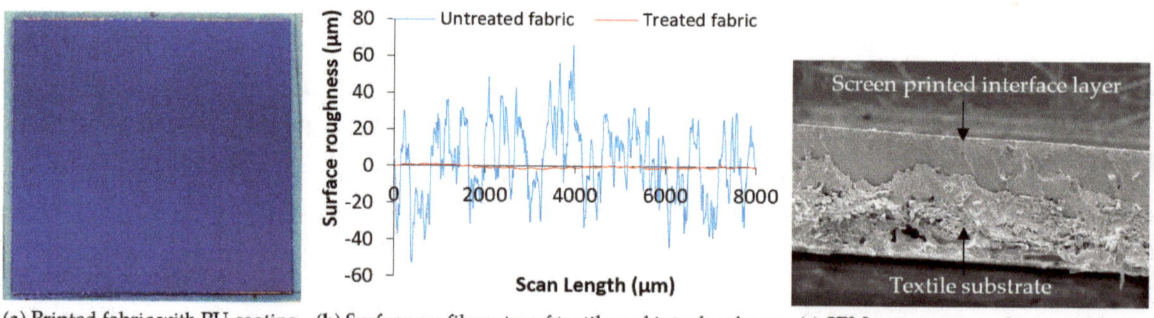

(**a**) Printed fabric with PU coating (**b**) Surface profilometry of textile and interface layer (**c**) SEM cross-section of printed fabric

Figure 2. Morphology and surface topology of printed fabric.

The printed fabric was metallized with 500 nm thick copper film using thermal evaporation deposition. A 10 nm thick chromium film was initially thermally deposited on the PU film to improve the adhesion of the copper film, as shown by the tape test in Figure 3.

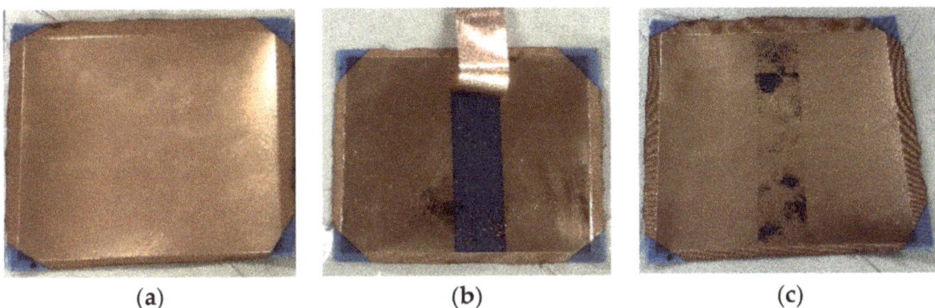

Figure 3. (**a**) thermally evaporated copper on interface layer; (**b**) adhesion of copper without and (**c**) with chromium layer after tape-test.

2.2. Evaluation of Solvents

To investigate the viability of the lift-off and etch processes in enabling microcircuit patterning on the interface layer, 10 by 85 mm samples of the untreated textile (i.e., without PU) and the PU-coated fabrics were prepared. The samples were immersed in the solvents listed in Table 1 between 5 min and 20 min to determine if the fabrics and PU coating would survive the solvents and their processing times.

2.3. Patterning Proccesses—Photolithography and Etching

To pattern the fabric, a 6 µm-thick negative photoresist, AZ2070, was spin-coated and baked at 110 °C for 3 min before and after UV exposure through a mask containing the different patterns with feature sizes, line width and spacing ranging from 10 µm to 800 µm as shown in Figure 1. The exposed resist was developed for 75 s in AZ726 developer solution, rinsed in de-ionized water and etched for 10 s in ferric chloride ($FeCl_3$) solution. Figure 4 shows the etched sample and the pattern resolutions achieved.

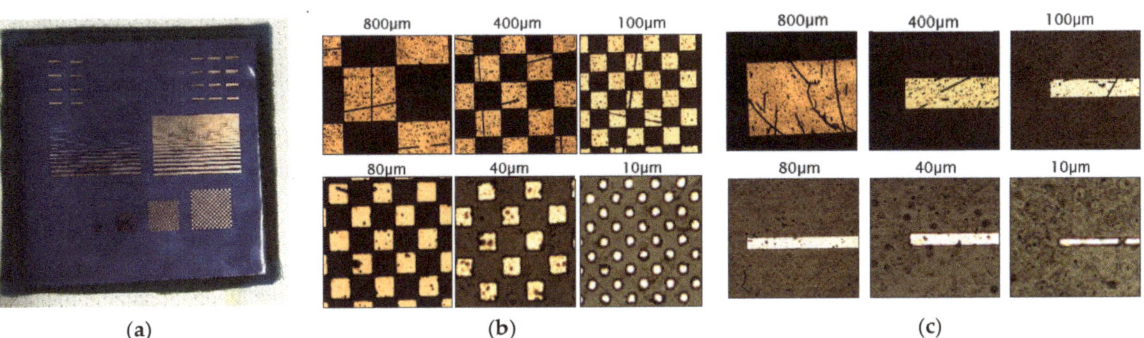

Figure 4. (**a**) Etched copper patterns on PU-coated fabrics; (**b**) feature quality of copper squares of different sizes; (**c**) resolution of different line width resolution.

3. Results and Discussion

Results comparing the various solvents demonstrate that they attack the PU coating, and this is especially noticeable in the lift-off process. N-Methyl-2-pyrrolidone (NMP), for example, severely degrades the PU coating after 20 min, as shown in Figure 5. The solvents for the etch process had minimal curling effect on the PU coating; hence, it was chosen for this work. For the wet stages of the lift-off and etching methods to be practical for micropatterning e-textiles, polymer friendly and gentle solvents are still required.

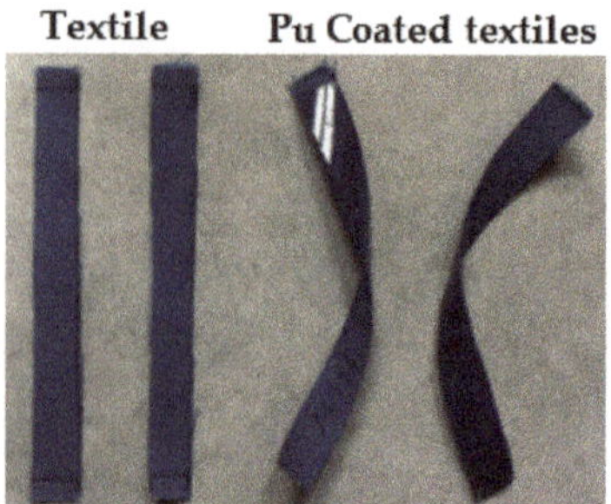

(a) Acetone for 5 mins

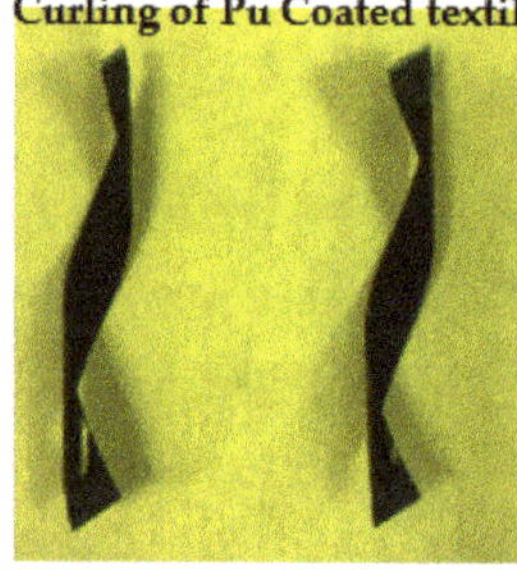

(b) Developer solutions for 5 mins

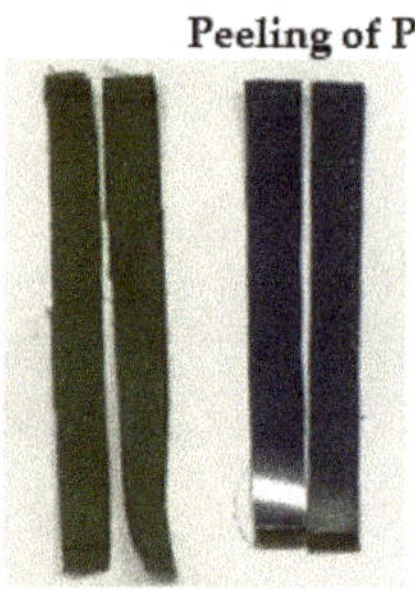

(c) $FeCl_3$ for 5 mins

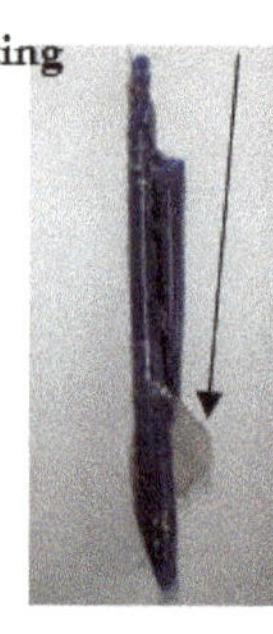

(d) NMP for 20 mins

Figure 5. Effect of solvent on screen-printed PU coating on fabric after immersion.

Figure 4 indicates that feature sizes down to 40 µm can be reliably etched on the PU-coated fabric, with linewidths and spacings of 10 µm also clearly defined. The yield and feature quality remain inadequate due to the poor adhesion of the copper film onto the interface layer, as indicated by the cracks in the patterns. The poor definition of the 10 µm copper squares further suggests that optimization of the fabrication process is required for higher resolutions.

4. Conclusions

This paper shows that microfabrication processes can be used in tandem with traditional screen-printing processes to reduce and improve form-factor and reliability of printed e-textiles, respectively. Feature sizes down to 10 µm have been realized on the PU-coated textile with this hybrid process. To achieve high yield and fine microcircuits, the defects and defect areas on the printed PU interface layer must be minimized. Furthermore, the adhesion of the thermally deposited films on the printed PU layer will need to improve to enhance reliability. Future work will also explore dry etching to mitigate the effect of solvents during fabrication.

Author Contributions: Conceptualization, A.K., M.G. and S.B.; methodology, A.K. and M.G.; design, investigation and formal analysis, A.K. and M.G.; writing—A.K. and M.G.; supervision, A.K. and S.B. All authors have read and agreed to the published version of the manuscript.

Funding: This research received no external funding.

Institutional Review Board Statement: Not applicable.

Informed Consent Statement: Not applicable.

Data Availability Statement: Not applicable.

Conflicts of Interest: The authors declare no conflict of interest.

References

1. Komolafe, A. Reliability and Interconnections for Printed Circuits on Fabrics. Ph.D. Thesis, University of Southampton, Southampton, UK, 2016.
2. Ohiri, K.A.; Pyles, C.O.; Hamilton, L.H.; Baker, M.M.; McGuire, M.T.; Nguyen, E.Q.; Currano, L.J. E-textile based modular sEMG suit for large area level of effort analysis. *Sci. Rep.* **2022**, *12*, 9650. [CrossRef] [PubMed]
3. Dils, C.; Kalas, D.; Reboun, J.; Suchy, S.; Soukup, R.; Moravcova, D.; Schneider-Ramelow, M. Interconnecting embroidered hybrid conductive yarns by ultrasonic plastic welding for e-textiles. *Textile Res. J.* **2022**, *92*, 4501–4520.
4. Stanley, J.; Hunt, J.A.; Kunovski, P.; Wei, Y. Novel Interposer for Modular Electronic Textiles: Enabling Detachable Connections between Flexible Electronics and Conductive Textiles. *IEEE Sens. Lett.* **2022**, *6*, 1–4. [CrossRef]

5. Zysset, C.; Cherenack, K.; Kinkeldei, T.; Tröster, G. Weaving integrated circuits into textiles. In Proceedings of the International Symposium on Wearable Computers (ISWC) 2010, Seoul, Republic of Korea, 10–13 October 2010; pp. 1–8.
6. Komolafe, A.; Torah, R.; Wei, Y.; Nunes-Matos, H.; Li, M.; Hardy, D.; Beeby, S. Integrating flexible filament circuits for e-textile applications. *Adv. Mater. Technol.* **2019**, *4*, 1900176. [CrossRef]

Proceeding Paper

Miniature Flexible Reprogrammable Microcontroller Circuits for E-Textiles †

Tom Greig [1,*], Kai Yang [2] and Russel Torah [1]

1 School of Electronics and Computer Science, University of Southampton, Southampton SO17 1BJ, UK
2 Winchester School of Art, University of Southampton, Southampton SO23 8DL, UK
* Correspondence: tg8g16@soton.ac.uk

† Presented at the 4th International Conference on the Challenges, Opportunities, Innovations and Applications in Electronic Textiles, Nottingham, UK, 8–10 November 2022.

Abstract: An e-textile system was developed, allowing USB reprogramming of miniature, flexible, integrated microcontroller circuits which allows for easier development of complex and configurable e-textile circuits. This prototype consisted of a series of five exposed pads on the edge of the PCB and a corresponding clip connector. Mounted onto the clip are a micro-USB port and necessary additional components to facilitate USB programming meaning that no additional components are required on the microcontroller board thus increasing flexibility. This system has the potential to make software development and reconfiguration of the e-textile easier while the small size and flexibility of the connector allow improved textile integration. This work provides a platform for future e-textile system development and increases the operational lifetime, thus reducing waste due to product obsolescence.

Keywords: e-textile; embedded microcontroller; e-waste; sustainability; flexible electronics

Citation: Greig, T.; Yang, K.; Torah, R. Miniature Flexible Reprogrammable Microcontroller Circuits for E-Textiles. *Eng. Proc.* **2023**, *30*, 15. https://doi.org/10.3390/engproc2023030015

Academic Editor: Steve Beeby

Published: 2 February 2023

1. Introduction

Microcontrollers are vital components in many e-textile devices [1,2]. Their re-programmability and wide range of peripheral functions means that they can fulfil the digital processing requirements of almost any small electronic product while their small size makes them possible to include in e-textile devices.

The ability to be re-programmed is key to a microcontroller's utility, and most provide some means of uploading new programs while in situ, for example, AVRs' SPI-based "ICSP" protocol [3]. However, such systems typically only work on one brand of microcontroller and require specialised programming circuits to use. The connectors required to use these systems also occupy a large area: the pin header needed to connect the ATMEL ICE programmer to a QFN ATtiny occupies 4 times the area and 12 times the height of the chip itself, see Figure 1.

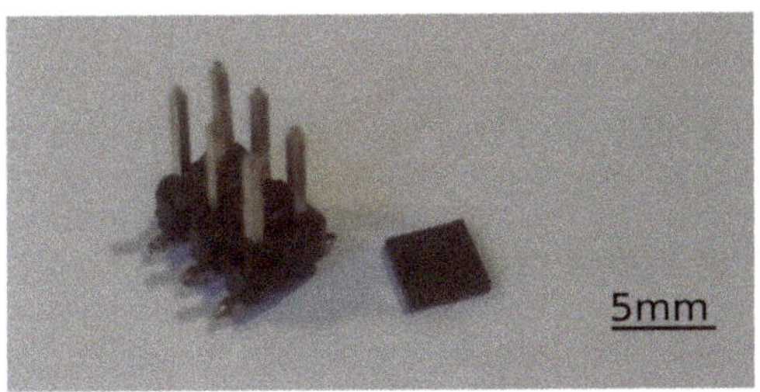

Figure 1. A standard 2.54 mm, 6-pin programming header (**left**) compared to the ATTiny85 microcontroller it programs (**right**).

The first problem can be solved using a USB bootloader. This is a small program which is used to load new software via a USB connection.

However, even micro-USB connectors are relatively large components in the context of e-textiles and restrict the textile integration of a microcontroller circuit. A potential solution to this is to use an edge connector. These consist of a series of exposed pads near the edge of the circuit board. The board itself is then inserted into a receptacle with contacts arranged to connect with the pads. However, existing solutions are for typically thicker (0.3 mm) flat flexible cables (FFC), therefore connections can be difficult and unreliable.

2. Design

The system developed here uses five, 1 mm pitch, pads which need only protrude a few millimetres from the body of a flexible circuit board connect to an external clip. The five connections are used for power, ground, positive and inverted USB data and a button-operated reset line which prompts the microcontroller to run its bootloader. The clip contains all the additional components needed for USB programming: the micro-USB connector, a 3.3 V voltage regulator, a reset button and several passive components. A diagram of this system is shown in Figure 2.

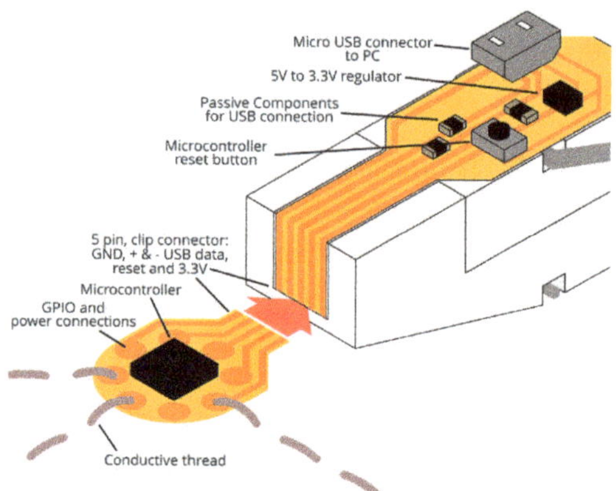

Figure 2. Design of the microcontroller and programmer. Placing the components needed for USB on the programmer clip means that only the microcontroller's IC needs to be integrated into the textile.

The contacts on both sides are tinned with solder to prevent corrosion This system was tested using an ATTiny85 microcontroller [3] with the micronucleus bootloader [4]. Both the flexible microcontroller circuit board and the programming clip's PCB were made using a standard photolithographic etching process described previously in [5].

The contacts on both sides are tinned with a thin layer of solder to prevent corrosion and to raise the contact point slightly, making the connection more reliable.

3. Applications

The test implementation was incorporated into both woven (Figure 3, top left) and stretchable, knitted fabrics (Figure 3, right) by couching the conductive thread soldered to the general purpose input/output (GPIO) pins for the controller. Another version was made by inserting the device and its connecting wires into woven pockets in a custom-made fabric (Figure 3, bottom left).

Because of its small size, the impact on the flexibility and stretchability of the fabric is minor. This is a major improvement over existing prototyping boards designed for e-textile which are much larger and used rigid PCBs (Figure 4).

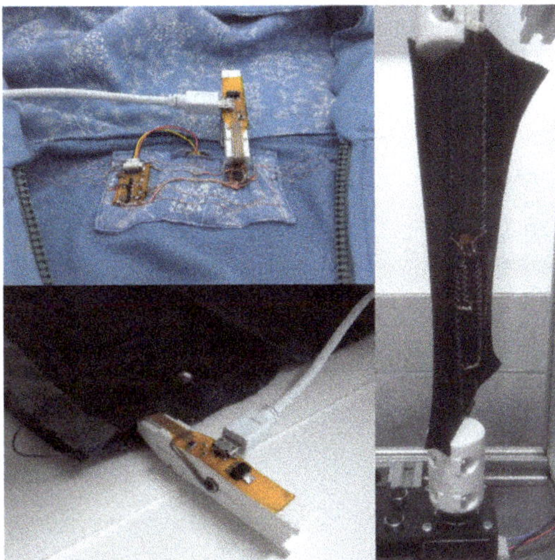

Figure 3. Microcontroller circuit integrated into the collar of a garment (**top left**), a stretchable knitted fabric (**right**), and a custom woven textile (**bottom left**).

Figure 4. Existing prototyping boards designed for e-textiles, left to right: an Arduino Lilypad (diameter 50 mm [6]), an Adafruit Gemma (diameter 28 mm [7]) and this work (10 × 12.5 mm).

This programming system can easily be adapted to other microcontrollers which are reprogrammable via USB.

4. Conclusions

This work presents an easy-to-use programmer, compatible with many different types of microcontrollers. It occupies significantly less circuit board space than existing commercial equivalents, the prototype displayed here is 80% smaller than an Adafruit Gemma and requires no additional components on the microcontroller board itself.

This is important in e-textile applications where a large size or additional rigid components can compromise the textile's properties of comfort and flexibility.

The small connector size and minor impact on integration mean that, when moving beyond the prototyping stage, the connector footprint may not need to be removed, and if it is, only a small change to the circuit layout is needed. The initial demonstrators also show that this methodology of flexible microcontroller integration and flexible connection point works with both couching into existing garment structures and integration during weaving.

Author Contributions: Conceptualization, investigation, writing—original draft preparation, T.G.; supervision, writing—review and editing, K.Y. and R.T. All authors have read and agreed to the published version of the manuscript.

Funding: This research was part-funded by EPSRC grant number EP/S001654/1, "Advanced e-textiles for wearable therapeutics" and part-funded by the EU H2020 programme, grant number WEARPLEX—825339—wearplex.soton.ac.uk.

Data Availability Statement: No new data was created or analysed in the study.

Conflicts of Interest: The authors declare no conflict of interest.

References

1. Castano, L.M.; Flatau, A.B. Smart fabric sensors and e-textile technologies: A review. *Smart Mater. Struct.* **2014**, *23*, 053001. [CrossRef]
2. Komolafe, A.; Zaghari, B.; Torah, R.; Weddell, A.S.; Khanbareh, H.; Tsikriteas, Z.M.; Vousden, M.; Wagih, M.; Jurado, U.T.; Beeby, S.; et al. E-Textile Technology Review–From Materials to Application. *IEEE Access* **2021**, *9*, 97152–97179. [CrossRef]
3. Datasheet for ATtiny85 Microcontroller from Microchip. Available online: https://www.microchip.com/en-us/product/ATtiny85 (accessed on 2 December 2022).
4. Micronucleus Bootloader Software for ATtiny85 USB Programming. Available online: https://www.github.com/micronucleus/micronucleus (accessed on 2 December 2022).
5. Komolafe, A.; Torah, R.; Wei, Y.; Nunes-Matos, H.; Li, M.; Hardy, D.; Dias, T.; Tudor, M.; Beeby, S. Integrating Flexible Filament Circuits for E-Textile Applications. *Adv. Mater. Technol.* **2019**, *4*, 1900176. [CrossRef]
6. Arduino Lilypad USB Documentation. Available online: https://docs.arduino.cc/retired/boards/lilypad-arduino-usb (accessed on 5 December 2022).
7. Adafruit Gemma v2 Product Page. Available online: https://www.adafruit.com/product/1222 (accessed on 5 December 2022).

Proceeding Paper

Design and Test of E-Textiles for Stroke Rehabilitation †

Meijing Liu [1], Tyler Ward [1], Odina Keim [2], Yuanyuan Yin [2], Paul Taylor [3], John Tudor [1] and Kai Yang [2,*]

1 Electronics and Computer Science, University of Southampton, Southampton SO17 IBJ, UK
2 Winchester School of Art, University of Southampton, Southampton SO23 8DL, UK
3 Odstock Medical Ltd., Salisbury SP2 8BJ, UK
* Correspondence: ky2e09@soton.ac.uk
† Presented at the 4th International Conference on the Challenges, Opportunities, Innovations and Applications in Electronic Textiles, Nottingham, UK, 8–10 November 2022.

Abstract: This work presents the design and test of an e-textile based functional electrical stimulation system for post-stroke upper limb rehabilitation. The prototype was tested on five stroke survivors to assess stimulation comfort, the stimulation intensity required to achieve hand opening, and ease of use. Wrist extension was measured using two inertial measurement units. The wearable e-textile prototype achieved similar stimulation comfort compared to high-quality hydrogel electrodes with a score difference of between 0 and 1. The stimulation intensity to achieve full hand opening was the same for the hydrogel electrodes and the e-textiles for all five participants. A second design based on a knitted sleeve has been assessed in terms of usability. Additional new designs have been proposed to improve the usability.

Keywords: electrode; functional electrical stimulation (FES); inertial measurement unit (IMU) stroke rehabilitation; e-textiles; healthcare

Citation: Liu, M.; Ward, T.; Keim, O.; Yin, Y.; Taylor, P.; Tudor, J.; Yang, K. Design and Test of E-Textiles for Stroke Rehabilitation. *Eng. Proc.* **2023**, *30*, 16. https://doi.org/10.3390/engproc2023030016

Academic Editors: Steve Beeby, Russel Torah and Theodore Hughes-Riley

Published: 6 February 2023

1. Introduction

Stroke occurs when there is a blockage or bleeding of the blood vessels affecting the supply of blood to the brain. There are 1.3 million stroke survivors in the UK [1] and stroke costs the UK National Health Service and wider society £26 billion per annum [2]. Over half of stroke survivors have weak arm/hand movement affecting their independence and quality of life. Functional Electrical Stimulation (FES) is a technology used for stroke rehabilitation. It applies a safe electrical impulse through electrodes placed on the skin to strengthen weak muscles and improve movement functions. FES has been used to exercise muscles and assist walking for people with mobility issues since the 1960s [3]. Systematic reviews with meta-analysis have concluded that FES improves the ability to perform activities [4–6]. Existing FES products are difficult to set-up by stroke survivors without help from their carers or healthcare professionals which significantly constrains usage. Our previous work has received positive feedback regarding wearable e-textile FES for home-based stroke rehabilitation [7]. This work presents the test results of a fabric electrode based wearable FES in terms of user comfort, stimulation intensity, and functional movement (wrist extension for hand opening) on five stroke survivors (ethics approval ID: University of Southampton ERGO 70296). Ease of use has been assessed. A second knitted design has been assessed and additional new designs to improve the usability have been proposed for future study.

2. Materials and Methods

2.1. Electrodes

Wearable electrodes (5 cm × 5 cm) were fabricated by stacking in turn: a non-woven fabric, conductive wires leading to a connector, a conductive carbon film and a carbon rubber electrode layer. Encapsulating the edges holds the entire assembly together. The electrode was attached to a fabric arm band as shown in Figure 1. The wearable electrodes were compared with commercial high quality hydrogel electrodes (PALS, Axelgaard, Fallbrook, CA, USA).

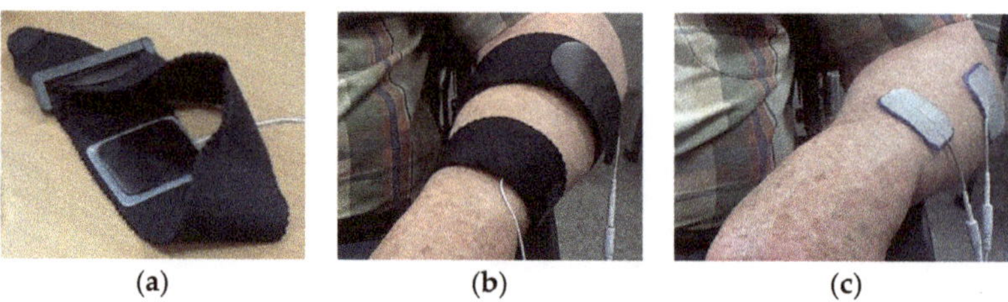

Figure 1. (**a**) fabric electrode arm band. (**b**) fabric electrode arm bands being worn by a stroke survivor. (**c**) hydrogel electrodes being worn by a stroke survivor.

2.2. FES

The OML Microstim 2V2 neuromuscular stimulator (Figure 2) was used in this study. Stimulation was set-up by a clinician to optimise the electrode positions and stimulation intensity. Water was sprayed on the electrodes and the skin before applying the fabric electrode to improve user comfort and stimulation effectiveness. Stimulation comfort was rated in a scale from 0 to 10 with 0 is the most comfortable and 10 being very painful. The stimulation intensity was recorded.

Figure 2. OML Microstim 2V2.

2.3. Inertial Measurement Unit (IMU) Sensors for Wrist Bending Measurement

The wrist angle was measured using a pair of BNO055 IMU sensor breakouts from Adafruit (Figure 3) which reports the absolute angular position of the sensor. The IMUs were attached to the user by using Velcro straps with one attached to the hand and the other to the arm. An Arduino Micro was used to gather the data from the sensors and sent it to the computer recording the data. The computer then calculated the wrist angle by taking the difference between the two absolute positions from the sensors.

Figure 3. BNO055 IMU sensor.

2.4. Usability Test

Participants were asked to put on and take off both the hydrogel electrodes and fabric electrode arm bands to assess their capability of using the products independently.

3. Test Results and Discussion

Five participants were recruited for the testing. Age: 56–76, Years of stroke: 3–17 years. Genders: 1 female and 4 males.

3.1. Stimulation Comfort

All participants reported the same or similar stimulation comfort with only 0 to 1 score difference between the two electrode types as shown in Table 1. One participant (P1) reported the fabric electrode was more comfortable than the hydrogel electrodes. One participant (P4) reported the same comfort scale. The other three reported the hydrogel electrodes were more comfortable than fabric electrodes. No pain sensation was reported.

Table 1. Stimulation comfort scale results by participant.

Participant	Hydrogel Electrode	Fabric Electrode
P1	3	2
P2	2–3	3
P3	5	6
P4	5	5
P5	2–3	3–4

3.2. Stimulation Intensity

There was no difference in the stimulation level required to achieve full hand opening between the two types of electrodes (Table 2). The required stimulation levels vary from 40 mA to 50 mA. This indicates the fabric electrodes were as effective as the hydrogel electrodes in generating a functional movement.

Table 2. Stimulation intensity required to achieve hand opening.

Participant	Hydrogel Electrode	Fabric Electrode
P1	50 mA	50 mA
P2	40 mA	40 mA
P3	40 mA	40 mA
P4	50 mA	50 mA
P5	45 mA	45 mA

3.3. Wrist Bending Measurement

Wrist movement was measured during the stimulation. All participants achieved a similar movement for the two types of electrodes. Figure 4 is a representative example of the wrist bending angle for the two electrode types.

3.4. Usability

Researchers observed that it was a challenging task for participants to peel off the hydrogel electrodes from the protective plastic film because the hydrogel electrodes are very sticky. It was even more challenging for them to take out the electrodes from the sealed bag and put them back after use. With the electrode arm band, although all participants were able to put it on and take it off independently, they found it challenging to keep the electrode in place because it moved around before it was fastened and secured in place. In addition, it was challenging to put the Velcro hook fastener through a loop. All participants were able to spray water on the electrodes and the arm using a single hand.

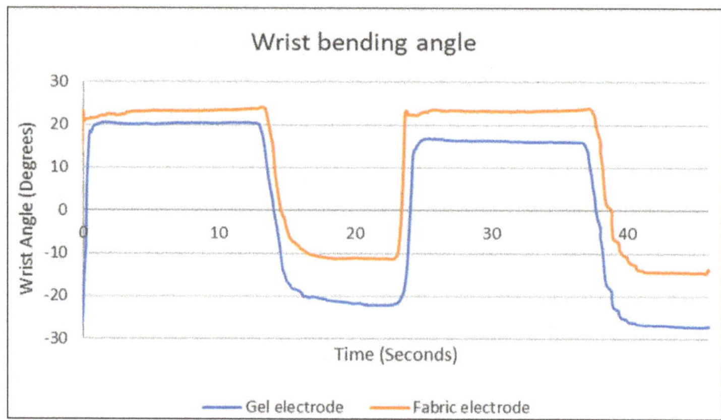

Figure 4. Wrist bending angle.

3.5. Second Design: Knitted Sleeve

Two pairs of electrodes were printed on a knitted fabric made of wool and Lycra yarns to stimulate muscles for both hand extension and flexion (Figure 5). The electrode fabric was assembled to form a pull-on sleeve. It was noticed the electrode sleeve was difficult to put on or take off because of the strong friction between the electrodes inside the sleeve and the skin. Discussions with the stroke survivors have indicated that an open, or partially open structure, would allow the user to put on the electrode sleeve easily, then tighten the sleeve to ensure the electrodes and skin contact sufficiently.

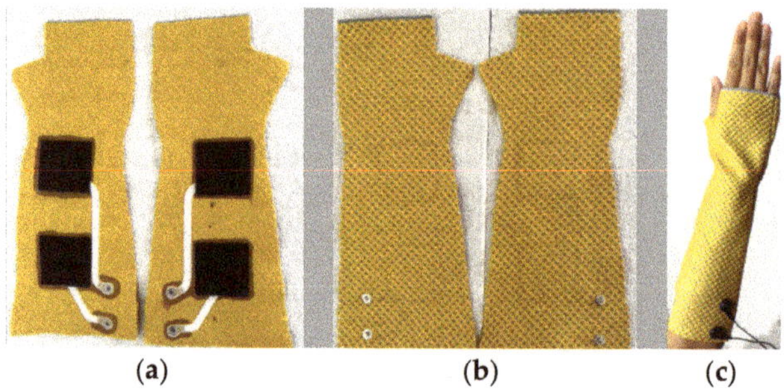

Figure 5. (**a**) electrodes printed on knitted fabric. (**b**) electrode fabric with snap buttons. (**c**) electrode sleeve worn on an arm and connected to a stimulator via snap buttons.

4. Conclusions and Future Work

This work has demonstrated the stimulation comfort and functional movement delivered by a wearable e-textile activated with the OML FES stimulator Microstim 2V2. The fabric electrode achieved similar stimulation performance while providing the advantage of being suitable for wearable application and offering a long service life. All participants were able to put on and take off the fabric electrode arm bands independently, but it is time consuming and a better design is required. A new design with electrodes printed on a knitted stretchable sleeve was investigated but it was difficult to put on and take off due to the friction between the electrodes inside the sleeve and the skin. New designs, required to improve the ease of use, will be addressed in future work.

Author Contributions: Conceptualization, M.L., J.T. and K.Y.; methodology, M.L., T.W., O.K., P.T. and K.Y.; software, T.W.; validation, M.L., Y.Y., P.T. and K.Y.; formal analysis, T.W. and K.Y.; investigation, M.L. and K.Y.; resources, P.T. and K.Y; writing—original draft preparation, M.L. and K.Y.; review and editing, All; funding acquisition, J.T. and K.Y. All authors have read and agreed to the published version of the manuscript.

Funding: This research was funded by the SIAH HEIF Research Stimulus Fund at the University of Southampton and the Medical Research Council (MRC) under grant number MR/N027841/1.

Institutional Review Board Statement: The study was approved by the Faculty of Arts and Humanities Ethics Committee at the University of Southampton on 21/02/2022 with a submission ID of ERGO 70296.

Informed Consent Statement: Informed consent was obtained from all subjects involved in the study.

Data Availability Statement: No new date was created.

Acknowledgments: The authors want to thank the stroke survivors at Different Strokes Southampton for attending the focus group studies and prototype testing.

Conflicts of Interest: The authors declare no conflict of interest.

References

1. Stroke Statistics, Stroke Association. Available online: https://www.stroke.org.uk/what-is-stroke/stroke-statistics (accessed on 4 December 2022).
2. Patel, A.; Berdunov, V.; Quayyum, Z.; King, D.; Knapp, M.; Wittenberg, R. Estimated societal costs of stroke in the UK based on a discrete event simulation. *Age Ageing* **2020**, *49*, 270–276. [CrossRef] [PubMed]
3. Vincent, G. Electrical Control of Partially Denervated Muscles. U.S. Patent 2737183A, 6 March 1956.
4. Veerbeek, J.M.; van Wegen, E.; van Peppen, R.; van der Wees, P.J.; Hendriks, E.; Rietberg, M.; Kwakkel, G. What Is the Evidence for Physical Therapy Poststroke? A Systematic Review and Meta-Analysis. *PLoS ONE* **2014**, *9*, e87987.
5. Howlett, O.A.; Lannin, N.A.; Ada, L.; McKinstry, C. Functional Electrical Stimulation Improves Activity after Stroke: A Systematic Review With Meta-Analysis. *Arch. Phys. Med. Rehabil.* **2015**, *96*, 934–943. [CrossRef] [PubMed]
6. Sun, J.; Yan, F.; Liu, A.; Liu, T.; Wang, H. Electrical Stimulation of the Motor Cortex or Paretic Muscles Improves Strength Production in Stroke Patients: A Systematic Review and Meta-Analysis. *J. Inj. Funct. Rehabil.* **2020**, *13*, 171–179. [CrossRef] [PubMed]
7. Liu, M.; Ward, T.; Keim, O.; Yin, Y.; Tudor, M.; Yang, K. Smart textile with integrated functional electrical stimulation and movement sensor for hand exercise. In Proceedings of the Annual Conference of the International Functional Electrical Stimulation Society (IFESS), RehabWeek, Rotterdam, The Netherlands, 25–29 July 2022.

Proceeding Paper

Wearable Preventive Pressure Ulcer System Using Embroidered Textile Electrodes †

Ghada Elbarbari [1,*], Wedian Madian [1], Barakat Mahmoud [2] and Bahira Gabr [1,3]

1 Apparel Design Management and Technology Department, Helwan University, Cairo 11795, Egypt
2 Plastic Surgery Department, Faculty of Medicine, Helwan University, Cairo 11795, Egypt
3 Faculty of Applied Arts, October 6 University, Cairo 12511, Egypt
* Correspondence: elbarbari.ghada@gmail.com

† Presented at the 4th International Conference on the Challenges, Opportunities, Innovations and Applications in Electronic Textiles, Nottingham, UK, 8–10 November 2022.

Abstract: Approximately 80 percent of patients with limited mobility experience pressure ulcers (PU). Electrical stimulation (ES) is an effective therapeutic approach for PU prevention and treatment. This study reports the design of a custom-made adaptive garment as a wearable preventive pressure ulcer system by using embroidered textile electrodes to induce electrical stimulation for the gluteal muscles. Eight electrodes were embroidered using a satin stitch on a 100% cotton knitted fabric, only the bobbin was loaded with conductive threads, and cotton/polyester thread was used for the top stitch. An ES-induced protocol of 1:4 s on–off was applied for 3 min, with 17 min rest periods.

Keywords: pressure ulcers; electrical stimulation; textile electrode; embroidery; conductive thread; adaptive apparel; patient garment

Citation: Elbarbari, G.; Madian, W.; Mahmoud, B.; Gabr, B. Wearable Preventive Pressure Ulcer System Using Embroidered Textile Electrodes. *Eng. Proc.* **2023**, *30*, 17. https://doi.org/10.3390/engproc2023030017

Academic Editors: Steve Beeby, Kai Yang, Russel Torah and Theodore Hughes-Riley

Published: 6 February 2023

1. Introduction

Pressure ulcers (PU) are a frequent complication encountered by doctors and patients and are a burden in terms of pain and treatment. They are localized tissue damage areas arising from excess pressure, shear, or friction [1]. They are common in patients with limited activity and mobility, such as paralyzed patients, surgical patients, and spinal cord injury (SCI) patients. Prevention has focused on support surfaces of cushions and mattresses to redistribute the interface pressure, but these passive methods do not address intrinsic risk factors such as tissue ischemia and decreased circulation [2].

On the other hand, electrical stimulation is an effective therapeutic prevention approach that decreases the interface pressure and activates the muscles to potentially reduce PU development [3]. The mechanism underlying the reduction is based on (i) changing blood flow that increases tissue oxygenation, which helps muscles to survive [4]; (ii) a decrease in tissue pressure caused by gluteal muscle contractions with the redistribution of pressure from ischial tuberosities (ITs) to the direction of the knees [5]; (iii) improved local circulation of muscle and skin and improved paralyzed muscle strength and mass [6]; (iv) muscle hypertrophy [7].

Patient clothing can contribute to objectification with a focus on disease and symptoms, and by doing so also contributes to the optimal treatment of ill health [8]. Therefore, textile-based electrodes can be ideal in medical high-tech applications; they are comfortable in a non-intrusive way and create a natural harmony with the patient's body. Textile electrodes, also known as textrodes, can be integrated into a garment using conventional textile production techniques by weaving, knitting, embroidery, etc. [9]. They are commonly used for biosignal monitoring as well as electrostimulation. However, embroidery offers advantageous characteristics of dimensional stability, rapid prototyping, better skin–electrode contact, good reproducibility, and flexibility with electrode designs [10].

2. Materials and Methods

2.1. Wearable Preventive Pressure Ulcer System Design

The electrical stimulation garment was designed to be adaptive and meet immobile patients' specific demands; it was designed with a seamless center crotch construction, to avoid bulky seams and to allow the electrode connection traces to run smoothly and freely without any interruption, as shown in Figure 1. The conductive thread, from Kitronik's (Kitronik's Electro-Fashion®, Nottingham, United Kingdom), was used for embroidering the electrodes using a conventional embroidery machine (JUKI LZ-271). A preliminary sample was made with one layer of 100% cotton knitted stretch fabric; the top stitch and bobbin were both loaded with conductive thread. A satin stitch was embroidered to a (4.5 × 4.5) filling electrode (12 stitches per cm/1 cm wide) and the connection trace was also created with a stain stitch (10 stitches per cm/4 mm wide), as shown in Figure 2. Unfortunately, running the machine was not easy with this specific thread; it was cut and jammed several times. As a result, the final electrodes were embroidered by simply filling the bobbin with the conductive thread.

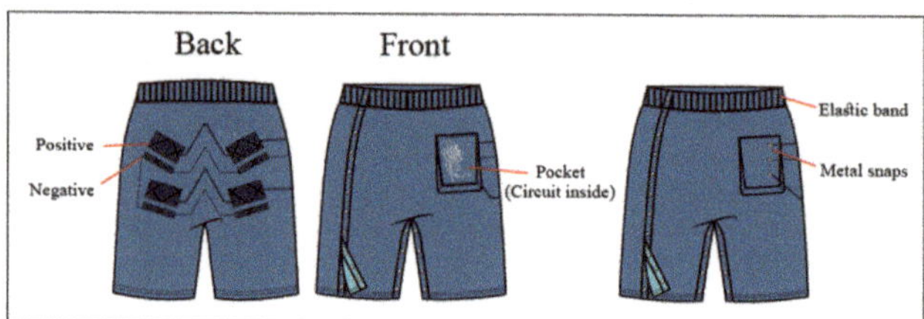

Figure 1. Illustration of the proposed electrical stimulation shorts.

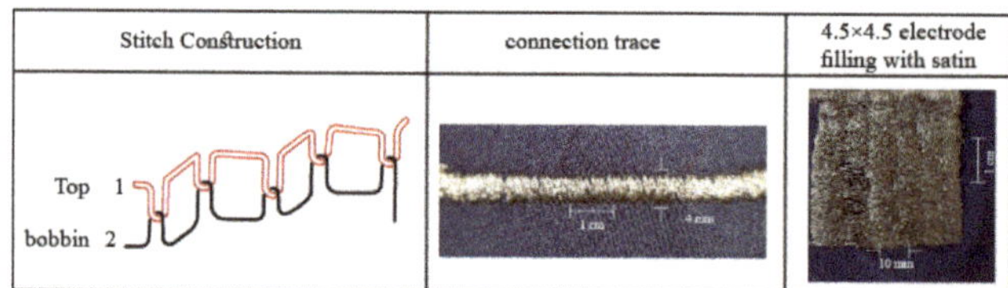

Figure 2. Close-up of stitched samples using a conventional embroidery machine.

The study conducted a custom-made circuit with an electrostimulation protocol of 1:4 s on–off that lasted for 3 min, with a 17 min rest period, as reported by Smit et al. [5] and shown in Figure 3; the gluteal muscle was activated for 1 s, and resting for 4 s resulted in better pressure relief and comfort without marked muscle fatigue. Parameters for the prevention of pressure ulcers that the circuit induced were set according to previous studies: frequency >20–50 Hz, amplitude 20–50 mA, and pulse width 64 to 600 μs [3].

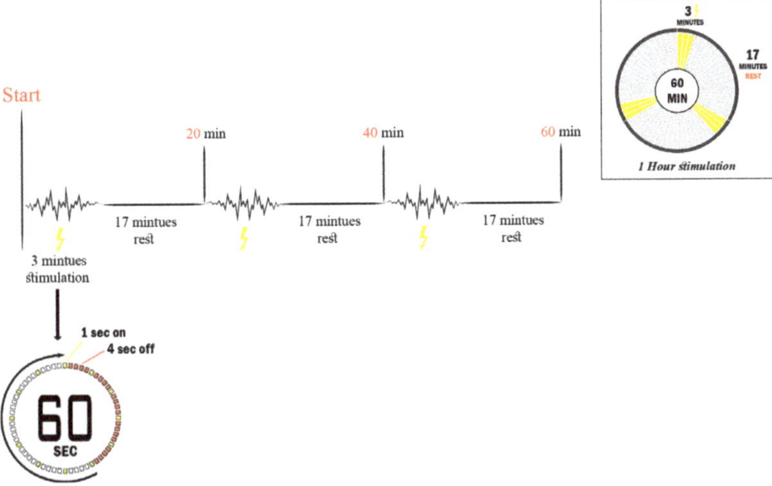

Figure 3. The 1 h stimulation protocol of 1:4 s on–off for 3 min, and 17 min rest [5].

2.2. Electrical Resistance of the Electrodes

Experiments were applied for the manufactured electrodes to evaluate their electrical performance. An AVO multimeter was used to measure the surface resistance of each electrode before and after washing, followed by the Martindale abrasion test. The change in resistance was compared with the number of washing cycles (up to 15 times with home laundry washing, interval of 5 times between each measurement, resulting in a total of 3 measurements), and the surface resistance was measured between cycles. The Martindale machine was used for up to 3000 rubs against raw wool, testing abrasion for the electrode test. Two prototypes were provided, one for each test.

3. Results and Discussion

3.1. Wearable Preventive Pressure Ulcer System Design

Regarding the embroidery process, the conductive thread showed superior performance when used on the bottom bobbin only of both the embroidery and sewing machine. Therefore, eight electrodes (4 identical positives 4 × 8 cm, 4 identical negatives 2 × 8 cm) were embroidered using a satin stitch, only the bobbin was loaded with conductive threads, and cotton/polyester thread was used for the top stitch. On the opposite side of the fabric, electrical traces were sewn with a lock stitch using the sewing machine; these traces were connected to metal snaps to be connected to the circuit, as shown in Figure 4.

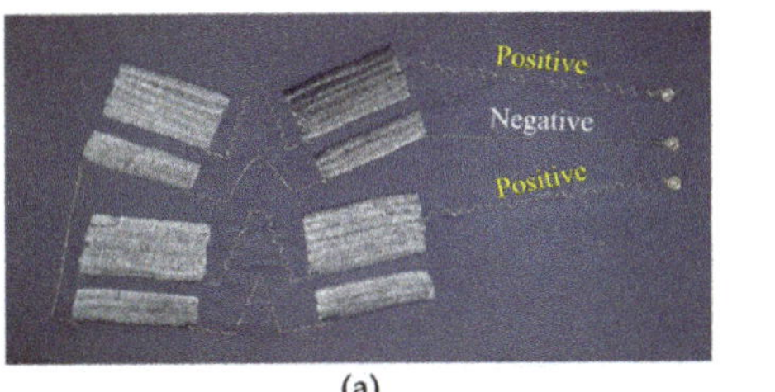

(a)

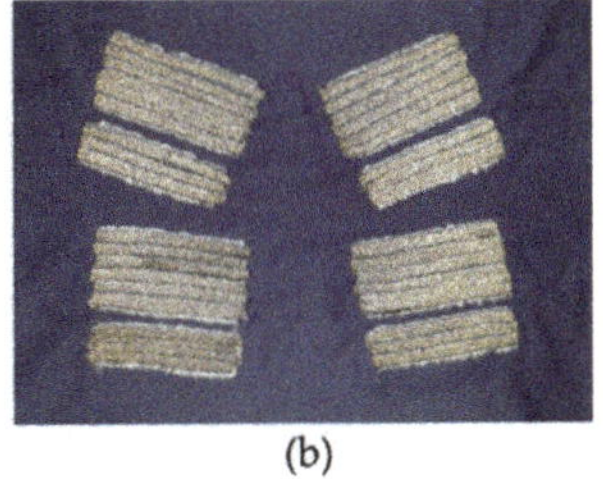

(b)

Figure 4. (**a**) The face of the fabric with cotton/poly; (**b**) the back of the fabric with the conductive thread.

3.2. Electrical Resistance of the Electrodes

The resistance of the embroidered textile electrodes was measured for each electrode and it was slightly changed after each washing cycle, as shown in Figure 5a. Similarly, the variation in resistance was almost stable before and after abrasion, as indicated in Figure 5b, with no observed visible damage in appearance. This indicated that the electrodes had the demanded flexibility and bendability for the purpose of electricity performance.

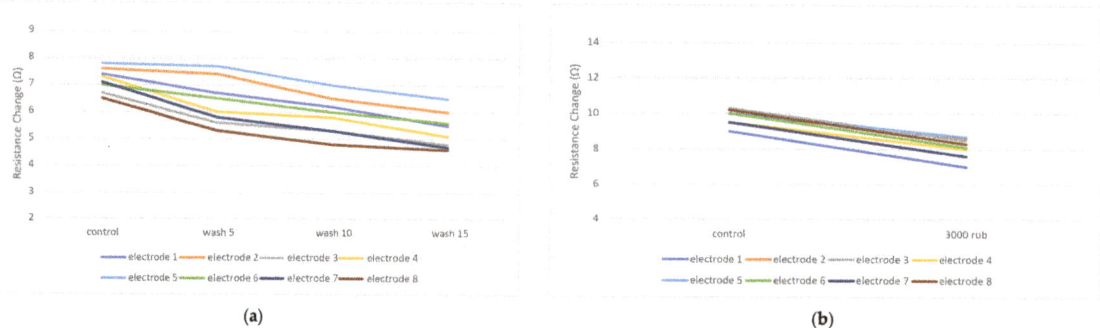

Figure 5. (**a**) Resistance value with washing cycle; (**b**) resistance change after abrasion.

4. Conclusions

Wearable technologies could offer promising tools as an alternative to physical therapy. This study designed a custom-made adaptive garment as a wearable preventive pressure ulcer system by using embroidered textile electrodes to induce electrical stimulation for the gluteal muscles. We developed a cable-free system that is comfortable, lightweight, flexible, easy to use for both caregiver staff and self-dressing patients, cost-effective for mass production and less time-consuming, and fit for purpose with ensured contact of the electrodes placed on the skin. Moreover, it eliminates the problems of conventional Ag/AgCl electrodes, which cause skin irritation and feelings of discomfort, cannot be washed, and are not hygienic.

Author Contributions: Conceptualization, G.E., W.M., B.M. and B.G.; methodology, software, validation, formal analysis, investigation, resources, data curation, writing—original draft preparation, visualization, and funding acquisition, G.E.; writing—review and editing, supervision, and project administration, W.M., B.M. and B.G. All authors have read and agreed to the published version of the manuscript.

Funding: This research received no external funding.

Institutional Review Board Statement: Not applicable.

Informed Consent Statement: Not applicable.

Data Availability Statement: Not applicable.

Acknowledgments: The authors would like to thank Ahmed Othman for the continued support and collaboration in creating the custom-made circuit; George Khalil and Hafez Hawas for helping with the technical and experimental tests; and Mahmoud Awad for the continued research support and collaboration in testing the proposed product.

Conflicts of Interest: The authors declare no conflict of interest.

References

1. Kandha Vadivu, P. Design and Development of Portable Support Surface and Multilayered Fabric Cover for Bed Sore Prevention. *Indian J. Surg.* **2015**, *77*, 576–582. [CrossRef] [PubMed]
2. Liu, L.Q.; Nicholson, G.P.; Knight, S.L.; Chelvarajah, R.; Gall, A.; Middleton, F.R.I.; Ferguson-Pell, M.W.; Craggs, M.D. Interface Pressure and Cutaneous Hemoglobin and Oxygenation Changes under Ischial Tuberosities during Sacral Nerve Root Stimulation in Spinal Cord Injury. *J. Rehabil. Res. Dev.* **2006**, *43*, 553. [CrossRef] [PubMed]
3. Liu, L.Q.; Moody, J.; Traynor, M.; Dyson, S.; Gall, A. A Systematic Review of Electrical Stimulation for Pressure Ulcer Prevention and Treatment in People with Spinal Cord Injuries. *J. Spinal Cord Med.* **2014**, *37*, 703–718. [CrossRef] [PubMed]
4. Gyawali, S.; Solis, L.; Chong, S.L.; Curtis, C.; Seres, P.; Kornelsen, I.; Thompson, R.; Mushahwar, V.K. Intermittent Electrical Stimulation Redistributes Pressure and Promotes Tissue Oxygenation in Loaded Muscles of Individuals with Spinal Cord Injury. *J. Appl. Physiol.* **2011**, *110*, 246–255. [CrossRef] [PubMed]
5. Smit, C.A.J.; Legemate, K.J.A.; de Koning, A.; de Groot, S.; Stolwijk-Swuste, J.M.; Janssen, T.W.J. Prolonged Electrical Stimulation-Induced Gluteal and Hamstring Muscle Activation and Sitting Pressure in Spinal Cord Injury: Effect of Duty Cycle. *J. Rehabil. Res. Dev.* **2013**, *50*, 1035–1046. [CrossRef] [PubMed]
6. Wijker, B.J.; de Groot, S.; van Dongen, J.M.; van Nassau, F.; Adriaansen, J.J.E.; Achterberg-Warmer, W.J.; Anema, J.R.; Riedstra, A.T.; van Tulder, M.W.; Janssen, T.W.J. Electrical Stimulation to Prevent Recurring Pressure Ulcers in Individuals with a Spinal Cord Injury Compared to Usual Care: The Spinal Cord Injury PREssure VOLTage (SCI PREVOLT) Study Protocol. *Trials* **2022**, *23*, 156. [CrossRef] [PubMed]
7. Thomaz, S.R.; Cipriano, G., Jr.; Formiga, M.F.; Fachin-Martins, E.; Cipriano, G.F.B.; Martins, W.R.; Cahalin, L.P. Effect of Electrical Stimulation on Muscle Atrophy and Spasticity in Patients with Spinal Cord Injury—A Systematic Review with Meta-Analysis. *Spinal Cord* **2019**, *57*, 258–266. [CrossRef] [PubMed]
8. Bergbom, I.; Pettersson, M.; Mattsson, E. Patient Clothing–Practical Solution or Means of Imposing Anonymity? *J. Hosp. Med. Manag.* **2017**, 3. [CrossRef]
9. An, X.; Stylios, G. A Hybrid Textile Electrode for Electrocardiogram (ECG) Measurement and Motion Tracking. *Materials* **2018**, *11*, 1887. [CrossRef]
10. Kim, H.; Kim, S.; Lim, D.; Jeong, W. Development and Characterization of Embroidery-Based Textile Electrodes for Surface EMG Detection. *Sensors* **2022**, *22*, 4746. [CrossRef] [PubMed]

Proceeding Paper

Development of the Smart Jacket Featured with Medical, Sports, and Defense Attributes using Conductive Thread and Thermoelectric Fabric †

Aman Ul Azam Khan [1,*], Aurghya Kumar Saha [1], Zarin Tasnim Bristy [2], Tasnima Tazrin [1], Abdul Baqui [1,‡] and Barshan Dev [1]

1 Department of Textile Engineering, BGMEA University of Fashion & Technology, Dhaka 1230, Bangladesh
2 Department of Fashion Design and Technology, BGMEA University of Fashion & Technology, Dhaka 1230, Bangladesh
* Correspondence: auazamkhantextile@gmail.com; Tel.: +880-1631-932267
† Presented at the 4th International Conference on the Challenges, Opportunities, Innovations and Applications in Electronic Textiles, Nottingham, UK, 8–10 November 2022.
‡ Project supervisor.

Abstract: The exigency of humans is boosting the necessity of Smart Textiles in this modern era. A decade ago, envisioning sophisticated outerwear with several uses were considered a challenge. This study aims to a jacket with 15 features; divided into 7 groups, including defense, sports, health, medical, women, and children safety mechanisms, 4 out of these 15 functions can be controlled by an Android app, "Smart Jacket BUFT". To avoid nonrenewable energy sources, solar power and energy harvesting technology to produce electricity from body heat and foot-powered energy were used, Smart jacket has embedded circuits and sensors alone with AD8232, MAX30100, NEO6m GPS, and ESP32 microcontrollers & voice and app-control. It is hopping that; his initial stage of growth and improvement will pave the way for subsequent activities.

Keywords: smart jacket; 15 features; android application; medical; defense; conductive thread; energy harvesting technology

Citation: Khan, A.U.A.; Saha, A.K.; Bristy, Z.T.; Tazrin, T.; Baqui, A.; Dev, B. Development of the Smart Jacket Featured with Medical, Sports, and Defense Attributes using Conductive Thread and Thermoelectric Fabric. *Eng. Proc.* **2023**, *30*, 18. https://doi.org/10.3390/engproc2023030018

Academic Editors: Steve Beeby, Kai Yang, Russel Torah and Theodore Hughes-Riley

Published: 7 February 2023

1. Introduction

The invention of textiles roughly 27,000 years ago could be regarded as the first time humans created a useful material [1]. Smart Textile can detect mechanical, thermal, magnetic, chemical, electrical, or other environmental variables and respond in a programmed manner (stimuli) [2]. Electronics are made from conductive threads and fabrics, whose limits and potential are determined by textile materials and production procedures [3]. In the late 1990s, MIT and the Georgia Institute of Technology conducted a series of studies on E-textiles in academia [4]. In the field of Textile Technology, it was hard to imagine the concept of Smart Jackets with all sorts of features, including health and medicinal functionalities, sports flexibilities, women's safety options, children's safety options, defense options, and a lot more, in just one jacket. The wearable electronic jacket is nowadays playing an important role in the medical world. New monitoring system support resources have been developed because of recent technological advances in mobile devices and wireless communications. A few years ago, the thought of a jacket that could seamlessly communicate and converse with a personal assistant and deliver practical solutions to everyday difficulties was simply a concept. This research paper brought the notion to life by designing a practical smart jacket with a total of fifteen features. To assure the jacket's flexibility and efficiency, we developed our own conductive thread and fabric flexible circuit. The Wearable Motherboard™ envisioned clothing that could monitor important signals discretely [5]. Cooseman et al. described a wireless charging garment with a patient monitoring system [6]. The use of carbon nanotubes to convert cotton thread for use in E-textiles was described in a 2008 study [7]. MyHeart, a 2009 EU (FP 6)-funded project, created ECG- and breathing-sensitive smart textiles [8]. For energy textile solution, in 2007, Qin et al. described energy harvesting utilizing Piezoelectric Zinc Oxide nanowires grown around textile fibers [9]. Design Research Lab at the Berlin University of the Arts and Telekom Innovation Laboratories designed the Smart Maintenance Jacket, which uses wearable technology for industrial maintenance work and predicts the future role of networked wearables in these contexts for telecommunications carriers [10]. Muhammad Arsalan et al. developed a next-generation tactical system for monitoring military health data. The suit's internal CPU receives and sends sensor data using LoRaWAN technology. The report analyzes the suit's system design and functionality [11]. Anurag Sharma et al. developed an IoT-based smart jacket in 2022 with ECG, heart rate, body temperature, air temperature, and oximeter sensors [12]. In the same year, Manibabu A and colleagues designed a Smart Thermal Jacket with Wearable Sensors using IoT [13]. In addition, Paolo Visconti et al. demonstrated a smart garment, in 2022, that monitors environmental factors and vital signs to monitor workers in hazardous industries [14]. The University of Engineering and Technology, Pakistan designed a wearable solar energy harvesting jacket for vital health monitoring systems in the current year [15]. Dilber Uzun Ozsahin et al. used multiple devices in one smart jacket [16].

2. Objectives

1. Design a multifunctional smart jacket from scratch.
2. To construct an energy-efficient jacket and to save non-renewable energy.
3. Utilize both hazardous and non-hazardous garbage, as well as clothes to decrease waste and diminish the effect on the environment.
4. Add Electrical Components to Textile Clothing.

3. Materials & Methods

3.1. Materials

The jacket was made with 100% polyester fabric of 300 GSM.

Trims: Plastic Zipper, Metal Zipper; Sewing Thread: 20/2, 40/3 and Fabric Glue was also used. Conductive thread and fabric flexible circuits maintain conductivity and add flexibility. To make it eco-friendly and cost-effective, waste materials were utilized. Besides, developed thermoelectric fabric was used which body heat into electrical energy. Table 1 shows the waste materials used in Smart Jacket.

Table 1. Waste Material Used in Smart Jacket.

Sl	Material	Form	Waste
1	Liquid Carbon Paste	Old dry cell battery	Hazardous wastage
2	Aluminum Foil	Old mobile phone battery	non-hazardous wastage
3	Copper foil	Old mobile phone battery	non-hazardous wastage
4	Empty Matchbox	Matchbox	General Wastage
5	Cutting Fabric	Apparel Industry	Apparel Wastage (Pre-Consumed)
6	Plastic bottle's crock	Plastic Wastage	Minimum solid waste
7	Synthetic Rubber	Old Rubber Gloves	General Waste

3.2. Methodology

Figure 1 shows Jacket's methodology according to units. The smart jacket's heart and brain work simultaneously. It collects data from NEO6M GPS module, MAX30100 Oxi-Pulse sensor, and AD8232 sensor circuit. An Android app was developed for Smart Jacket's brain and heart. Before starting the app, ESP32 was connected to a mobile hotspot. The NEO6M GPS module sent a signal to satellites. GPS started when satellites sent data to NEO6M GPS module. Firebase displays the user's location, heart rate, oxygen level, and

ECG via a mobile app after receiving ESP32 data. (Figure 2b) Despite using two 1000 mAh 3.7 V DC batteries for a total voltage of 7.2 V and an LM7805 MOSFET as a 5.0 V voltage regulator, they were all turned on by 5.0 V. This 7.2-volt battery was charged by a 5.0-volt, two-piece, 1-watt solar panel on the back of the jacket.

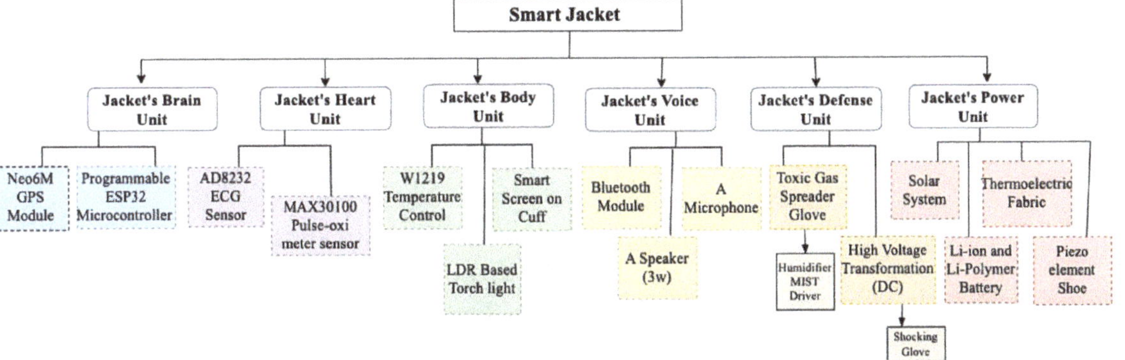

Figure 1. Flowchart of Jacket's methodology according to units.

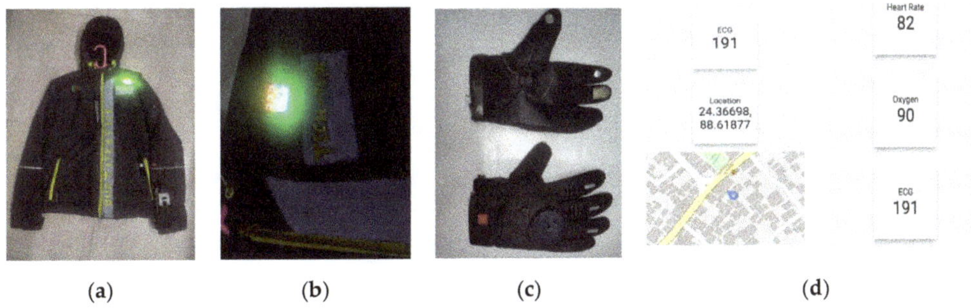

Figure 2. (**a**) Final Smart Jacket (**b**) Dual Display of automatic heat-warm system; (**c**) Jacket Defense Unit (**d**) The Android application's data of the smart jacket.

The body unit was equipped with an automated heating system, automatic light on/off, a smart screen on the cuff, and a device charging system. Automatic heating used a W1219 temperature controller circuit and an NTC sensor module. It was a heating/cooling controller. First, a W1219 circuit was connected to a 14.8-volt power bank charged by a shoe's piezo element generated electricity. Three heating pads were on W1219. As the user walked, mechanical energy was transferred to piezo components and converted to electrical energy. The potential difference's energy was then converted with a DC-to-DC converter and sent to a 14.8-volt power source. Nichrome wire and cotton made a heating pad formed a solenoid. The heating pad warmed. W1219 has two screens. The two displays showed the current and set temperatures. The user could set the temperature to their personal preferences. Moreover, the automated light on/off system constructed by a flexible fabric circuit with a resistor, transistor, and LDR. The battery was 3.7 V, 800 mAh DC which was charged by thermoelectric fabric. Thermoelectric fabric was formed from electronic and textile waste. When light-dependent resistors detect more light, their resistance rises, deactivating the circuit. When the LDR sensed no light, its resistance dropped, triggering the circuit. This turned on the jacket's hoodie LEDs.

Electronic device charging system contained by a 3.7 volt, 3600 mAh Lithium-ion battery with a voltage boost module. A 5-volt, 1-watt solar panel charged this battery. A USB voltage booster was attached to the battery in the welt pocket. This study featured a smart screen on the cuff of the smart jacket which was created with a fabric flexible circuit using Arduino pro mini, HC-05 Bluetooth module, 0.96″ OLED display, 3.7-volt 2800 mAh battery (which can be charged by1 watt, 5.0-volt solar panel), and polyester fabric and conductive thread. Figure 2c exhibits the methods of the smart screen the on cuff.

The jacket's voice unit consisted of an old speaker PCB circuit, 3.7 volts and 1000 mAh a lithium-ion battery (which can be charged by 1 watt, 5.0-volt solar panel), a speaker (3 w) and a microphone. A Bluetooth circuit and an mp3 circuit were combined to from this circuit whose connection was formed with the smartphone upon activation. Personal assistants such as Google, Siri, and Alexa can be linked into smartphones.

In the jacket's right-glove there is a shocking element. For that, a high-voltage transformer whose input was 3.7 to 6 volts and output were 400,000 DC volt, a metal shank button, a 3.7-volt DC lithium-ion battery, and a 5.0-volt 1-watt mini solar panel. A high voltage transformer was connected to shank button and placed on the glove's two-finger. The glove had to be waterproof and for extra safety we used synthetic rubber on the shank button area. Upon pressing the switch shank button, 400,000 DC volts were produced, allowing them to defend themselves easily. The left glove has a toxic gas spreader system with an ultrasonic MIST module, a switch, and a 1000 mAh, 3.7-volt lithium polymer battery, which was charged by a 5-volt 1-watt solar panel. For MIST, we made a liquid reserve tank from plastic waste. MIST module had a piezo atomizer disc. When activated by battery, the MIST module main circuit of IC sent data to the disc, which turned liquid into gas. As a self-defense glove, chloroform ($CHCl_3$) was used. So, users can easily spread toxic gas ($CHCl_3$) for defense or protection purposes.

4. Results

All this jacket's functionalities were executed excellently. Figure 2a–c show the practical visualization of the jacket. An Android mobile application was developed and given the name "SMART JACKET BUFT" for these four functions (GPS with emergency switch, Oximeter, pulse meter, ECG). Figure 2d shows the Android application of the smart jacket. These features can be accessed by multiple users who are using the same mobile applications as the wearer. the power unit uses solar system (which covers most functions), piezoelectric shoe, and thermoelectric fabric (which is still under testing). However, during this research we were unable to find any kind of flexible battery, so we used tiny-sized lithium polymer and lithium-ion batteries.

Formatting of Mathematical Components

For the solar system,

Estimate voltage and current from the mini solar panel [17].

$$\text{Charging time} = \frac{\text{battery capacity (watt hours)}}{\text{solar power (in watt)}} \times 2 \tag{1}$$

On the back part of this jacket are five small solar panels that serve as the power output's charging mechanism. Each solar panel has dimensions of 90 mm by 90 mm (l × w) and voltage of 5.00 volts. Table 2 estimated charging time according to unit.

Table 2. Estimated charging time according to unit.

Features (According to Unit)	No of Solar	Num of Battery	Battery Voltage (V)	Battery Amphere (mAh)	Charging Time (Hour)
Heart and Brain unit	2	2	3.7 × 2 = 7.4	1000	14.8
Shocking glove (Defense unit) + Electrical device charging system (Body Unit)	1	2	3.7 × 2 = 7.4	3600	53.28
Toxic gas spreader glove (defense unit) + Voice unit	1	2	3.7 × 2 = 7.4	1000	14.8
Smart screen (Body Unit)	1	1	3.7	1000	7.4

5. Features

The jacket includes 15 features.

1. GPS with emergency switch: a Custom-built Android-app to determine GPS-enabled user's real time location, which will notify if the user presses the emergency-button.
2. Oximeter: this jacket's oxygen meter allows the wearer to check their level via a mobile app.
3. Pulse-meter: this jacket's pulse meter can be monitored via app.
4. ECG: it has an ECG-measurement system that can be accessed via app.
5. Smart screen on cuff: Bluetooth lets users connect to the jacket's smart screen that displays the user's phone's time, SMS, and email.
6. Automatic Heat Warm System: this instantly warms the wearer. This feature maintains user-selected temperature.
7. Automatic light on-off system: in the dark, two powerful LED torches on the jacket's hood turn on. The light turns off when the user goes from dark to light.
8. Charging system: It's 'belt'pocket has a charging mechanism for mobile devices.
9. Shocking glove: it is for women's self-defense. The jacket's 400,000 DC voltage will likely shock an attacker.
10. Toxic gas spreader glove: it is used for self-defense. Jacket users can defend by spreading narcotic gases like chloroform.
11. Fashion feature: jacket had many pockets, such as tablet, waterproof, hidden, pen holder pockets etc.
12. RFID: by placing the phone in the RFID pocket, it enters 'Flight-mode'.
13. Solar panel: solar panel on the back of the jacket powers most of its functions.
14. Thermoelectric fabric: it was used in the smart jacket for experimental purpose.
15. Piezo element shoe: this jacket has a pair of shoes with piezo elements.

6. Discussion & Conclusions

6.1. Discussion

This study sought to improve the adaptability of textiles and electronic components. Conductive thread was utilized to connect both flexible fabric circuits and fabric-printed circuit boards. To manufacture the most popular conductive thread, however, carbon nanotubes, graphene, and silver were utilized, which was relatively expensive [18]. For this study, a conductive thread was developed with a meager resistance (0.2 Ω/cm to 0.0164 Ω/cm). Therefore, many circuits in one jacket will not be a problem. The jacket research encompassed all human requirements. This garment can be employed to construct seven jackets in the future. Medical, Military, Women's, Children's, Tech-Tex, Cyberia Survival, and Sports Jackets. E-textile research also requires Flexible Nanochips and Nano electric particles which were not available in BD. In addition, the required chemical composition for this experiment was not available in BD. Moreover, the financial crisis was the most crucial truth about this research. Despite showing that entire jackets can be developed with Fabric PCB shown in Figure 3 and Fabric Flexible Circuits. A piezo-shoe jacket attachment was also created for charging reasons. This feature aimed to construct wireless electrical charging technology; however, it failed due to lack of resources, time, and knowledge. This investigation was concluded despite a few flaws. Any researcher working on e-textiles or intelligent textiles might benefit from this study.

Figure 3. Fabric PCB.

6.2. Future Outlook

Regarding the characteristics, this investigation discovered further revolutionary concepts. Some of the jacket's attributes may not have existed because of economic slump. Among other innovations, there was power generated from human hair, a small, flexible tab on apparel, a 20× magnifying camera, a poisonous gas detector, an air oxygen level meter, and a button projector. In addition, there were several medical features, such as an automatic blood pressure measuring system, a body temperature meter, and galvanic skin resistance, and others. In addition to defensive functions, such as a web shooter, laser burning system on the glove, CO_2 spreading system, heads-up translating display, 100× magnification camera glasses, and several others, the smart jacket would be outfitted with an assortment of extra functionality. This investigation also offers some original ideas for the defense industry. As additions to robotics, two micro robots resembling Mr. Doctor and Mr. Commander were proposed. These intelligent micro robots will be concealed within the flap pockets. Importantly, jacket advancements include anti-bacterial fabric, ECG electrode fabric, fire-prevention fabric, and bulletproof fabric. Energy-harvesting textiles, such as solar fabric, would boost the value of clothing. The next generation of smart jackets would be powered by artificial intelligence. In the medical and military industries, this jacket improvement would be incredibly useful. Micro- and nanotechnology would enhance the jacket's performance, reduce its weight, and make it more comfortable.

Author Contributions: Conceptualization, A.U.A.K.; A.K.S.; A.B.; Z.T.B.; methodology, A.U.A.K.; A.B.; software, A.K.S.; validation, A.B.; Z.T.B.; T.T.; formal analysis, A.B.; A.U.A.K.; investigation, A.B.; B.D.; resources, A.K.S.; A.U.A.K.; data curation, A.U.A.K.; Z.T.B.; A.K.S.; writing—original draft preparation Z.T.B.; A.U.A.K.; writing—review and editing, A.B.; Z.T.B.; B.D.; visualization, A.U.A.K.; Z.T.B.; supervision, A.B.; project administration, A.B.; B.D. All authors have read and agreed to the published version of the manuscript.

Funding: The authors would like to acknowledge financial support from the BGMEA University of Fashion and Technology, Dhaka 1230, Bangladesh.

Institutional Review Board Statement: Not applicable.

Informed Consent Statement: Not applicable.

Data Availability Statement: The data presented in this study are available within the article and there is presented in every graph. There is no more data apart from the presented.

Acknowledgments: Erin Jahan Meem, Department of Textile Engineering, BGMEA University of Fashion & Technology. Taslima Ahmed Tamanna, Department of Textile Engineering, BGMEA University of Fashion & Technology.

Conflicts of Interest: The authors declare no conflict of interest.

References

1. Adovasio, J.M.; Soffer, O.; Klíma, B. Upper Palaeolithic fibre technology: Interlaced woven finds from Pavlov I, Czech Republic, c. 26,000 years ago. *Antiquity* **1996**, *70*, 526–534. [CrossRef]
2. Syduzzaman, M.D.; Patwary, S.U.; Farhana, K.; Ahmed, S. Smart Textiles and Nano-Technology: A General Overview. *J. Text. Sci. Eng.* **2015**, *5*, 181. [CrossRef]
3. Stewart, R. Cords and Chords: Exploring the Role of E-Textiles in Computational Audio. *Front. ICT* **2019**, *6*, 2. [CrossRef]
4. Post, E.R.; Orth, M. Smart fabric, or 'wearable clothing'. In Proceedings of the Digest of Papers. First International Symposium on Wearable Computers, Cambridge, MA, USA, 13–14 October 1997; pp. 167–168. [CrossRef]
5. Gopalsamy, C.; Park, S.; Rajamanickam, R.; Jayaraman, W.S. The Wearable Motherboard?: The first generation of adaptive and responsive textile structures (ARTS) for medical applications. *Virtual Real.* **1999**, *4*, 152–168. [CrossRef]
6. Coosemans, J.; Hermans, B.; Puers, R. Integrating wireless ECG monitoring in textiles. *Sens. Actuators A Phys.* **2006**, *130–131*, 48–53. [CrossRef]
7. Shim, B.S.; Chen, W.; Doty, C.; Xu, C.; Kotov, N.A. Smart Electronic Yarns and Wearable Fabrics for Human Biomonitoring made by Carbon Nanotube Coating with Polyelectrolytes. *Nano Lett.* **2008**, *8*, 4151–4157. [CrossRef] [PubMed]
8. Harris, M.; Habetha, J. The MyHeart project: A framework for personal health care applications. In Proceedings of the 2007 Computers in Cardiology, Durham, NC, USA, 30 September–3 October 2007; pp. 137–140. [CrossRef]
9. Qin, Y.; Wang, X.; Wang, Z.L. Microfibre–nanowire hybrid structure for energy scavenging. *Nature* **2008**, *451*, 809–813. [CrossRef] [PubMed]
10. Greinke, B.; Guetl, N.; Wittmann, D.; Pflug, C.; Schubert, J.; Helmut, V.; Bitzer, H.-W.; Bredies, K.; Joost, G. Interactive workwear: Smart maintenance jacket. In Proceedings of the 2016 ACM International Joint Conference on Pervasive and Ubiquitous Computing: Adjunct, Heidelberg, Germany, 12–16 September 2016; pp. 470–475. [CrossRef]
11. Arsalan, M.; Musani, A.A.; Ailia, S.A.; Baig, N.; Shaikh, E.M.K. Military Uniform for Health Analytics for Field Intelligent Zone (MUHAFIZ) Protecting the ones that protect our land. In Proceedings of the 2018 2nd International Conference on Smart Sensors and Application (ICSSA), Kuching, Malaysia, 24–26 July 2018; pp. 64–68. [CrossRef]
12. Sharma, A.; Sharma, A.; Paul, M.R. An IoT-Based Smart Jacket for Health Monitoring with Real-Time Feedback. In *Harnessing the Internet of Things (IoT) for a Hyper-Connected Smart World*, 1st ed.; Apple Academic Press: Boca Raton, FL, USA, 2022; pp. 63–90. [CrossRef]
13. Manibabu, A.; Lakshmi, M.V.; Netish, V.P.; Padmanaban, A. Smart Thermal Jacket with Wearable Sensors using IoT. In Proceedings of the 2022 6th International Conference on Computing Methodologies and Communication (ICCMC), Erode, India, 29–31 March 2022; pp. 439–444. [CrossRef]
14. Visconti, P.; de Fazio, R.; Velazquez, R.; Al-Naami, B.; Ghavifekr, A.A. Wearable sensing smart solutions for workers' remote control in health-risk activities. In Proceedings of the 2022 8th International Conference on Control, Instrumentation and Automation (ICCIA), Tehran, Iran, 2–3 March 2022; pp. 1–5. [CrossRef]
15. Khan, A.S.; Khan, F.U. A Wearable Solar Energy Harvesting Based Jacket with Maximum Power Point Tracking for Vital Health Monitoring Systems. *IEEE Access* **2022**, *10*, 119475–119495. [CrossRef]
16. Ozsahin, D.U.; Almoqayad, A.S.; Ghader, A.; Alkahlout, H.; Idoko, J.B.; Duwa, B.B.; Ozsahin, I. Development of smart jacket for disc. In *Modern Practical Healthcare Issues in Biomedical Instrumentation*; Academic Press: Cambridge, MA, USA, 2021; pp. 31–46. [CrossRef]
17. Available online: https://blog.voltaicsystems.com/estimating-battery-charge-time-from-solar/ (accessed on 1 February 2023).
18. Alagirusamy, R.; Textile Institute (Eds.) *Technical Textile Yarns: Industrial and Medical Applications*; CRC Press: Boca Raton, FL, USA, 2010.

Proceeding Paper

The Machine-Learning-Empowered Gesture Recognition Glove †

Jun Luo [1], Yuze Qian [1], Zhenyu Gao [2], Lei Zhang [2,*], Qinliang Zhuang [1,*] and Kun Zhang [1,*]

[1] Key Laboratory of Textile Science & Technology (Ministry of Education), College of Textiles, Donghua University, Shanghai 201620, China

[2] Engineering Research Center of Digitized Textile & Fashion Technology (Ministry of Education), College of Information Science & Technology, Donghua University, Shanghai 201620, China

* Correspondence: lei.zhang@dhu.edu.cn (L.Z.); qlzhuang@dhu.edu.cn (Q.Z.); kun.zhang@dhu.edu.cn (K.Z.)

† Presented at the 4th International Conference on the Challenges, Opportunities, Innovations and Applications in Electronic Textiles, Nottingham, UK, 8–10 November 2022.

Abstract: Recently, gesture recognition technology has attracted increasing attention because it provides another means of information exchange in some special occasions, especially for auditory impaired individuals. At present, the fusion of sensor signals and artificial intelligence algorithms is the mainstream trend of gesture recognition technology. Therefore, this article designs a machine-learning-empowered gesture recognition glove. We fabricate a flexible strain sensor with a sandwich structure, which has high sensitivity and good cycle stability. After the sensors are configured in the knitted gloves, the smart gloves can respond to different gestures. Additionally, according to the representation characteristics and recognition targets of sampled signal data, we explore a segmented processing method of dynamic gesture recognition based on Logit Adaboost algorithm. After classification training, the recognition accuracy of smart gloves can reach 97%.

Keywords: gesture recognition; flexible sensor; machine learning; Logit Adaboost algorithm

Citation: Luo, J.; Qian, Y.; Gao, Z.; Zhang, L.; Zhuang, Q.; Zhang, K. The Machine-Learning-Empowered Gesture Recognition Glove. *Eng. Proc.* **2023**, *30*, 19. https://doi.org/10.3390/engproc2023030019

Academic Editors: Steve Beeby, Kai Yang, Russel Torah and Theodore Hughes-Riley

Published: 21 February 2023

1. Introduction

As a method of nonverbal communication [1], standardized gestures play an important role in the transmission of message on specific occasions, such as communication between auditory-impaired individuals [2] and information interaction in AR/VR environments [3] or military fields [4]. However, without long-term systematic learning, it is difficult for most people to understand the meaning of gestures [5]. At present, gesture recognition technology offers a feasible solution to this communication barrier, which can effectively judge the specific movements of the hand.

Machine-vision-based image taking and processing has received extensive attention as a means of gesture recognition due to simple equipment requirements [6]. However, it is troubled by a series of problems in the process of application, including complex operation mechanism and variable lighting condition [7]. Accelerometers and gyroscopes are widely used in commercial gesture recognition devices, which can track the direction of the hand movement in three-dimensional space [8]. However, rigid materials and excessive volume limit the application scenarios of these devices, which bring inconvenience to users. Benefiting from the rapid advancement of flexible electronics and artificial intelligence, wearable flexible data gloves based on flexible sensors have been developed. Common flexible sensors for gesture recognition include piezoresistive sensors [9], capacitive sensors [10], piezoelectric sensors [11], triboelectric sensors [12] and EMG sensor arrays [13]. Through the electromechanical performance of flexible sensors, the deformation degree of the hand can be reflected correspondingly.

Furthermore, the classification of sensor signals and the extraction of gesture features are the key factors to determine the accuracy of gesture recognition. It generally requires intelligent algorithms to train sensor signals and generate appropriate classifiers [14]. However, due to the differences in gesture behavior patterns of individuals, the generalization ability of the trained classifiers is low if the sample capacity is limited [15]. A common approach is to increase the number of people participating in algorithm training, which is time- and energy-consuming [16]. Therefore, there is a new idea to remove the irrelevant factors of sensor signals as much as possible through special designs.

In this study, we report on the machine-learning-empowered gesture recognition glove. Flexible piezoresistive strain sensors with high sensitivity and cycle stability were configured in knitted gloves. Then, in order to enhance the generalization ability of gesture recognition model, a multi segment classification model of dynamic gesture recognition based on Logit AdaBoost algorithm was explored. Static signals collected at intervals of gesture changes are usually considered invalid signals because of unobvious characteristics and uncertain duration. So, the model distinguished static signals from effective dynamic signals due to finger movements. Additionally, subsequent training and classification was only related to dynamic signals, avoiding misjudgment caused by static signals. With sufficient training for the defined gestures, the recognition accuracy reached 97%, showing great classification performance.

2. Materials and Methods

2.1. Materials

Reduced graphene oxide (r-GO) was supplied by Qingdao University. Polydimethylsiloxane (PDMS) was purchased from Microflu Co., Ltd. (Changzhou, China). *N,N*-Dimethylformamide (DMF, 99.5%) was bought from Aladdin Co., Ltd. (Shanghai, China).

2.2. Fabrication of the Flexible Strain Sensor

The r-Go dispersion (6 wt%) was prepared by two steps. The reduced graphene oxide powders were firstly dispersed in *N,N*-Dimethylformamide by mechanical shearing for 1 h at 3500 r/min, and, secondly, it was necessary to remove air bubbles from the solution using ultrasonic for 5 min. Then, the solution was poured onto the PDMS substrate (the volume ratio of PDMS matrix to curing agent was 10:1 and the curing condition was 70 °C for 1 h) and heated at 50 °C for 3 h using a hot plate. After the solvent evaporates completely, there was a composite film containing the r-Go layer and PDMS substrate. Finally, two copper wires were adhered to both sides of the r-Go layer with silver paste, which used as the electrodes of the strain sensor. Additionally, PDMS was poured and cured on the top of the composite film, resulting in a sandwich structure. The size of the final sample is about 35 mm × 4 mm × 1 mm.

2.3. Characterization

A scanning electron microscope (SEM, S-4800, Hitachi, Japan) was used to characterize the surface morphology of the R-Go film. Additionally, in order to test the sensing performance of the strain sensors, the flexible strain sensors were fixed between the metallic clamps of a stretching machine (ZQ-CI701G, 1000 N, ZHIQU, Dongguan, China), and a digital source-meter (Keithley 2400, USA) was used to measure the resistance of the sensors continuously in the process of stretching. The stretching speed was maintained at 30 mm/min.

2.4. The Preparation of Data Gloves and Data Acquisition

The configuration of strain sensors in data glove is shown in Figure 1a. To monitor the bending degree of fingers, the sensors were attached to the corresponding finger joint position of knitted gloves. Figure 1b shows the gesture library of this article. When making these gestures, the resistance signal acquisition of sensors was realized by the data acquisition card (Ni-6211). In the detection circuit, sensors were connected in parallel, and divider resistors were connected in series with the sensor on each branch. Based on the differential mode of the data acquisition card, the voltage on both sides of the sensors were collected completely and displayed in a laptop program designed with LabVIEW. The resistance of the sensor can be converted by Ohm's law.

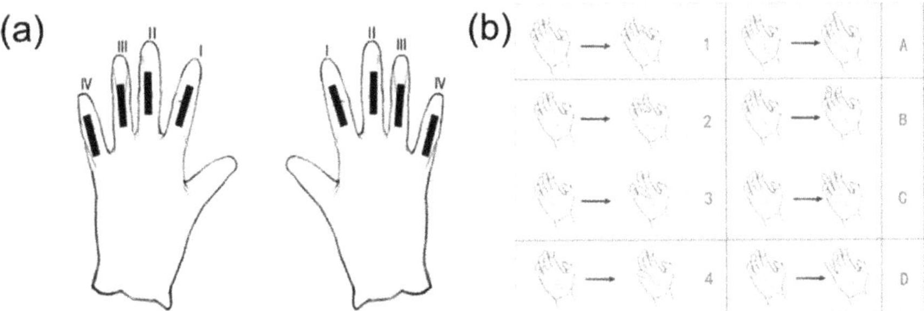

Figure 1. (**a**) Configuration of flexible sensor in data glove. (**b**) Gesture library.

2.5. The Design of Machine-Learning Algorithm

A flow diagram of the machine-learning process is shown in Figure 2. During the training process, the change in voltage due to predetermined gestures was firstly merged into a matrix as the input of mRMR algorithm to discard useless information and retain relevant signals, which improves the generalization ability of the model and reduces the complexity of the algorithm. Subsequently, the multi-segment classification model was applied to classify gestures by using the extracted features with Logit AdaBoost algorithm. Based on the training results of the previous weak classifier, the Logit AdaBoost algorithm evaluated and redefined the parameter weights to update the classifier. Additionally, the Logit AdaBoost algorithm was not sensitive to noise and outliers, which prevented the model from over fitting to a certain extent. The classification process included two steps. First, static signals and dynamic signals were effectively distinguished. Then, combined with the sample characteristics and distribution, the motion trajectory of the finger was further subdivided. Finally, 10-fold cross-validation method was used to evaluate the model. It can make full use of the samples in the training process, which avoids over fitting.

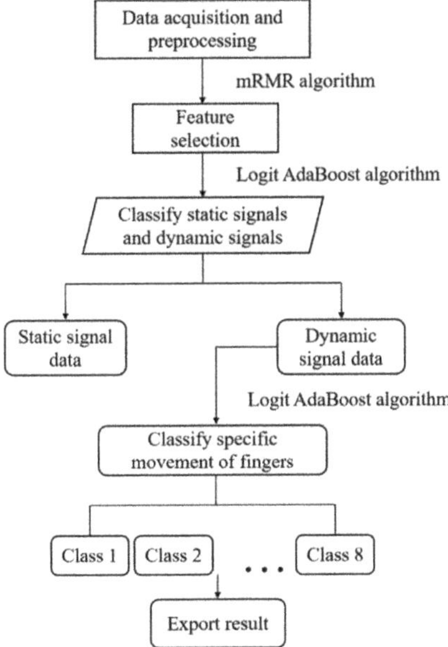

Figure 2. The process of a multi-segment classification model.

3. Results and Discussions

3.1. Morphology Characterization

As shown in Figure 3a, based on a sandwich structure composed of a r-GO layer and two PDMS layers, the overlapping state of graphene sheets can change during the deformation of the sensor and recover under the constraint of elastic material (PDMS), resulting in the electromechanical response of sensors.

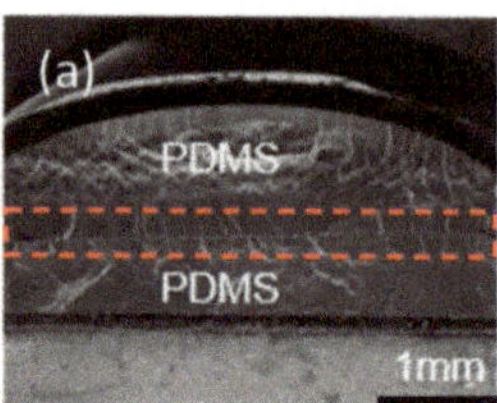

Figure 3. Topography characterization of the sensor. (**a**) Cross-section of the sensor. (**b**) SEM images of the r-GO film.

Figure 3b shows the that reduced graphene oxide is uniformly distributed in the r-GO layer, which is the basis of ensuring the stable electrical performance of the sensor.

3.2. Piezoresistive Sensing Response

Figure 4a shows the relationship between the sensor resistance and applied strains, from which the sensitivity of the strain sensor can be obtained. The sensitivity of the strain sensor can be calculated by the following equation:

$$GF = \frac{\Delta R / R_0}{\varepsilon} \quad (1)$$

$$\Delta R = R - R_0 \quad (2)$$

where ε is the strain of the strain sensor, R represents the resistance of sensor after stretching, R_0 is the resistance of sensors at ε = 0%. Figure 4a shows the strain sensor provides high sensitivity for detecting hand movements (GF~2.69 at strains below 12%, GF~5.05 at the strain ranges from 12% to 36%). Furthermore, the strain sensors also show a stability response over a large number of stretch–release cycles (above 1500 cycles), as shown in Figure 4b.

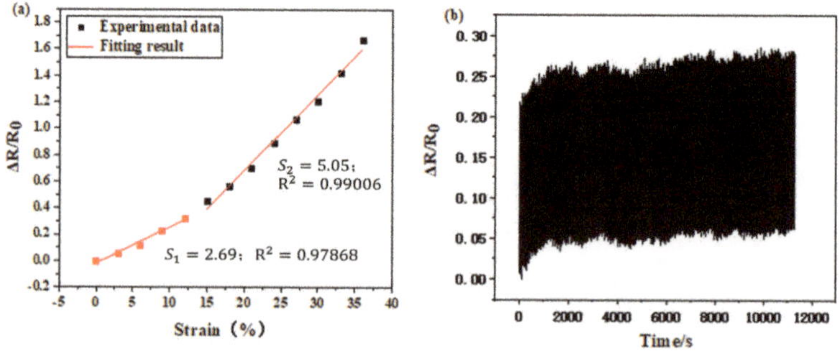

Figure 4. (**a**) The relationship between the sensor resistance and the applied strain. (**b**) Sensor stability under 1500 stretch—release cycles at 10% strain.

3.3. The Selection of Classification Algorithm and Feature

In order to find the data correlation, we extracted 10 characteristics as signal features, which are related to indicators of time domain, frequency domain and entropy. Then, we used the mRMR algorithm to check the importance of the features in the two stages. As shown in Table 1, waveform factors and skewness have the great correlation with the classification results of the first stage. Skewness and waviness are of significance for the classification of the second stage.

Table 1. Feature importance evaluation based on mRMR algorithm.

The First Stage		The Second Stage	
Features	**Importance Weight**	**Features**	**Importance Weight**
waveform factor	0.8679	skewness	0.5681
skewness	0.1321	kurtosis	0.4319
other features	10^{-16}–10^{-27}	other features	10^{-17}–10^{-25}

RUSBoost, Logit AdaBoost and Bagging are the common machine learning algorithms in data classification. Based on selected features through the mRMR algorithm, the classification results of the three algorithms on gesture data are shown in Figure 5. Area under curve (AUC) is a statistical concept. It represents the probability that the predicted value of positive cases is greater than the predicted value of negative cases. Whether it is the first stage or the second stage, the AUC of Logit AdaBoost algorithm is maximum. It shows that the model generated by Logit AdaBoost algorithm is most effective for gesture classification.

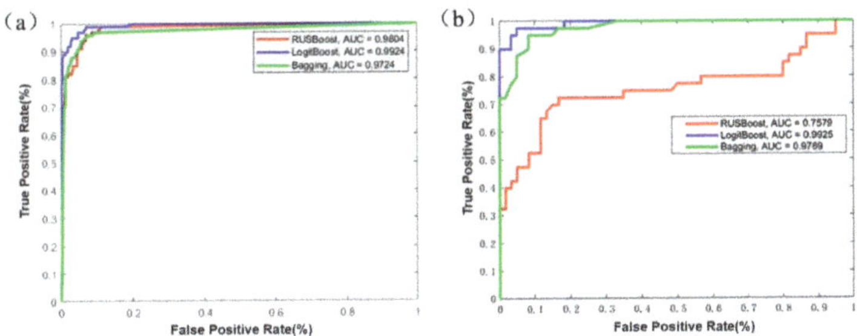

Figure 5. Comparison of model performance by AUC for RUSBoost, Logit AdaBoost, Bagging. (**a**) Performance of different algorithm models in the first stage. (**b**) Performance of different algorithm models in the second stage.

3.4. The Gesture Recognition Modeling

In the first stage, dynamic signals are separated from all samples for subsequent training. As shown in Figure 6a, the classification model can correctly classify most of the static gesture data and dynamic gesture data. Table 2 further shows the classification results of static signals and dynamic signals. Positive predictive value (PPV) is an index to evaluate the level of correct classification. The lower PPV of dynamic signals, the more sample capacity is lost in the second stage, resulting in unobvious and incomplete features as gesture output. Additionally, false discovery rate (FDR) represents the probability of misclassification, which affects the interference degree of data directly. High PPV level (96.9%) and low FDR level (3.1%) illustrate that effective gesture features are completely preserved.

The ROC curve after 10-fold cross-validation is shown in Figure 6b. Assuming that the dynamic signal is positive, the AUC value of the model in the first stage is 0.99 and the accuracy reaches 98.4%, indicating that the model can further eliminate invalid static signals.

Based on defined gestures, the movement of the hand is divided into the bending and straightening of fingers. Figure 6c and Table 3 shows the accuracy of classification remains at a high level. Additionally, Figure 6d illustrates the AUC value of the model is 0.97, which only has subtle decreases compared with the first stage. The good classification results of two stages indicates that the multi-segment classification model has the generalization ability for gesture recognition.

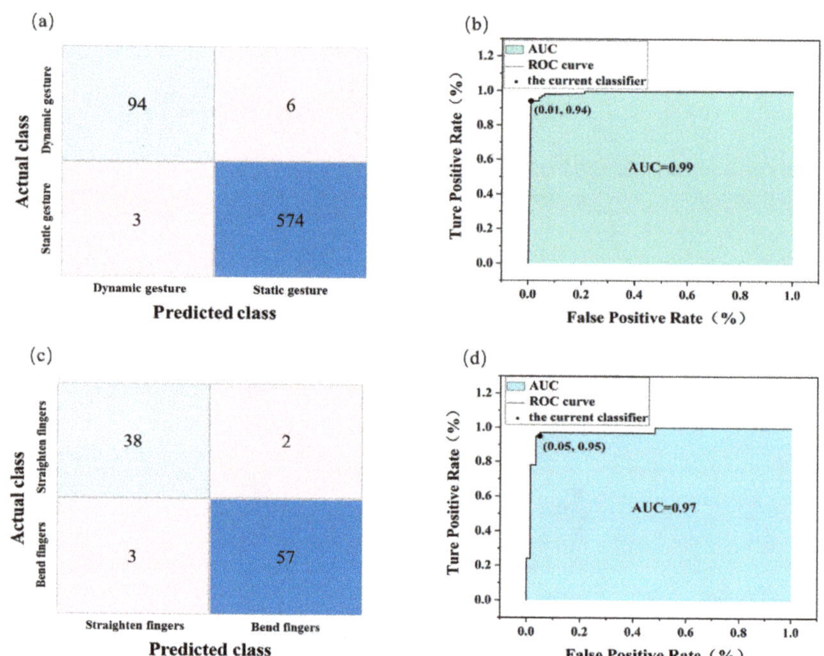

Figure 6. Test results for classification model of two phases. (**a**) Confusion matrix of dynamic gesture and static gesture. (**b**) ROC curve of classification model assuming dynamic gesture is positive. (**c**) Confusion matrix of straightening fingers and bending fingers. (**d**) ROC curve of classification model assuming straightening fingers is positive.

Table 2. Evaluation of classification model to distinguish static signals and dynamic signals.

Index	Static Signals	Dynamic Signals
True Positive Rate	99.3%	94.0%
False Positive Rate	0.7%	6.0%
Positive Predictive Value	98.7%	96.9%
False Discovery Rate	1.3%	3.1%

Table 3. Evaluation of classification model to judge finger motion trajectory.

Index	Bend Fingers	Straighten Fingers
True Positive Rate	95.0%	95.0%
False Positive Rate	5.0%	5.0%
Positive Predictive Value	96.6%	92.7%
False Discovery Rate	3.4%	7.3%

4. Conclusions

This article reported a type of machine-learning empowered gesture recognition glove. Combined with the electrical characteristics of flexible strain sensors and the signal processing ability of machine learning algorithms, data gloves judged the specific actions of hands accurately, which showed many commendable features, such as low weight and good comfortableness. At the same time, we paid attention to the interference of invalid static signals in the training and classification process. The model successfully extracted effective dynamic signals from all signals through multi-segment classification based on Logit AdaBoost algorithm. It is of significance to enhance the generalization ability of the recognition model and increase the accuracy of gesture recognition. However, not all irrelevant factors have been excluded for this model, which requires further research.

Author Contributions: Conceptualization, K.Z.; methodology, J.L., Y.Q. and Z.G.; writing—original draft preparation, J.L., Y.Q. and Z.G.; writing—review and editing, J.L., Y.Q., Z.G., L.Z., Q.Z. and K.Z.; supervision, L.Z., Q.Z. and K.Z.; funding acquisition, K.Z. All authors have read and agreed to the published version of the manuscript.

Funding: This research was funded by the Fundamental Research Funds for the Central Universities (2232022G-01 and 19D110106), the National Natural Science Foundation of China (NO. 51973034).

Institutional Review Board Statement: Not applicable.

Informed Consent Statement: Not applicable.

Data Availability Statement: The data presented in this study are available on request from the corresponding author.

Conflicts of Interest: The authors declare no conflict of interest.

References

1. Galván-Ruiz, J.; Travieso-González, C.M.; Tejera-Fettmilch, A.; Pinan-Roescher, A.; Esteban-Hernández, L.; Domínguez-Quintana, L. Perspective and Evolution of Gesture Recognition for Sign Language: A Review. *Sensors* **2020**, *20*, 3571. [CrossRef] [PubMed]
2. Pan, W.; Zhang, X.; Ye, Z. Attention-based sign language recognition network utilizing keyframe sampling and skeletal features. *IEEE Access* **2020**, *8*, 215592–215602. [CrossRef]
3. Tsironi, E.; Barros, P.; Weber, C.; Wermter, S. An analysis of convolutional long short-term memory recurrent neural networks for gesture recognition. *Neurocomputing* **2017**, *268*, 76–86. [CrossRef]
4. Saggio, G.; Cavrini, F.; Pinto, C.A. Recognition of arm-and-hand visual signals by means of SVM to increase aircraft security. *Stud. Comput. Intell.* **2017**, *669*, 444–461.
5. Xue, Q.; Li, X.; Wang, D.; Zhang, W. Deep forest-based monocular visual sign language recognition. *Appl. Sci.* **2019**, *9*, 1945. [CrossRef]
6. Weichert, F.; Bachmann, D.; Rudak, B.; Fisseler, D. Analysis of the Accuracy and Robustness of the Leap Motion Controller. *Sensors* **2013**, *13*, 6380–6393. [CrossRef] [PubMed]
7. Ameen, S.; Vadera, S. A convolutional neural network to classify American Sign Language fingerspelling from depth and colour images. *Expert Syst.* **2017**, *34*, e12197. [CrossRef]
8. Shukor, A.Z.; Miskon, M.F.; Jamaluddin, M.H.; bin Ali, F.; Asyraf, M.F.; bin Bahar, M.B. A new data glove approach for Malaysian sign language detection. *Procedia Comput. Sci.* **2015**, *76*, 60–67. [CrossRef]
9. Lee, B.G.; Lee, S.M. Smart wearable hand device for sign language interpretation system with sensors fusion. *IEEE Sens. J.* **2018**, *18*, 1224–1232. [CrossRef]
10. Abhishe, K.S.; Qubeley, L.C.F.; Ho, D. Glove-based hand gesture recognition sign language translator using capacitive touch sensor. In Proceedings of the 2016 IEEE International Conference on Electron Devices and Solid-State Circuits (EDSSC), Hong Kong, China, 3–5 August 2016; pp. 334–337.
11. Lecun, Y.; Bengio, Y.; Hinton, G. Deep learning. *Nature* **2015**, *521*, 436–444. [CrossRef]
12. Zhou, Z.; Chen, K.; Li, X.; Zhang, S.; Wu, Y.; Zhou, Y.; Meng, K.; Sun, C.; He, Q.; Fan, W.; et al. Sign-to-speech translation using machine-learning-assisted stretchable sensor arrays. *Nat. Electron.* **2020**, *3*, 571–578. [CrossRef]
13. Wu, J.; Sun, L.; Jafari, R. A wearable system for recognizing American sign language in real-time using IMU and surface EMG sensors. *IEEE J. Biomed. Health Inform.* **2016**, *20*, 1281–1290. [CrossRef] [PubMed]
14. Rivera-Acosta, M.; Ortega-Cisneros, S.; Rivera, J.; Sandoval-Ibarra, F. American sign language alphabet recognition using a neuromorphic sensor and an artifcial neural network. *Sensors* **2017**, *17*, 2176. [CrossRef] [PubMed]
15. Cai, Y.D.; Feng, K.Y.; Lu, W.C.; Chou, K.C. Using LogitBoost classifier to predict protein structural classes. *J. Theor. Biol.* **2006**, *238*, 172–176. [CrossRef] [PubMed]
16. Ho, C.Y.F.; Ling, B.W.K.; Deng, D.X.; Liu, Y. Tachycardias Classification via the Generalized Mean Frequency and Generalized Frequency Variance of Electrocardiograms. *Circuits Syst. Signal Process.* **2022**, *41*, 1207–1222. [CrossRef]

Proceeding Paper

Zinc-Ion Battery on a Polyester-Cotton Textile †

Sheng Yong [1,*], Nick Hillier [1,2] and Steven Beeby [1]

1 Smart Electronic Materials & System Research Group, Electronics and Computer Science, University of Southampton, Southampton SO17 1BJ, UK
2 Energy Storage and Its Applications Centre of Doctoral Training, University of Southampton, Southampton SO17 1BJ, UK
* Correspondence: sy1v16@soton.ac.uk; Tel.: +44-2380523119
† Presented at the 4th International Conference on the Challenges, Opportunities, Innovations and Applications in Electronic Textiles, Nottingham, UK, 8–10 November 2022.

Abstract: This work presents a simple, scalable and flexible zinc-ion secondary battery, fabricated on top of a textile substrate via standard fabrication processes. The proposed zinc-ion battery was fabricated on top of a polyester-cotton textile using solution-based processes and inexpensive cathode, anode and electrolyte materials. This battery achieved an area capacity of 19.1 μAh·cm^{-2} between 1.9 and 0.9 V.

Keywords: e-textile; flexible battery; zinc-ion battery

Citation: Yong, S.; Hillier, N.; Beeby, S. Zinc-Ion Battery on a Polyester-Cotton Textile. *Eng. Proc.* **2023**, *30*, 20. https://doi.org/10.3390/engproc2023030020

Academic Editors: Steve Beeby, Kai Yang, Russel Torah and Theodore Hughes-Riley

Published: 16 March 2023

1. Introduction

Wearable electronics also known as e-textile are the integration of flexible electrical devices in clothing and accessories. In these systems, sufficient electrical energy is required for devices such as sensor nodes [1], microprocessors and transceivers [2]. At present, such systems are typically powered using an external electrical connection, battery or supercapacitor [3] which require frequent replacement [4] and/or need to be physically removed before washing. Therefore, a flexible energy-storage device such as a zinc-ion battery (ZIB) integrated on top of or inside a textile is important for e-textile applications.

ZIBs are rechargeable secondary electrical energy-storage devices. In comparison to ordinary secondary batteries such as lithium-ion and aluminium air batteries. This type of battery demonstrates distinct advantages, such as improving device safety by not using aggressive or environmentally unfriendly electrolytes (in lithium-ion batteries) and can be fully sealed without the requirement of oxygen access such as in an aluminium air battery. In addition, the assembly of ZIBs can be performed in the air, offering reduced fabrication complexity. Previously Yong et al. [5] presented a zinc-ion secondary textile battery. The manganese oxide cathode, zinc anode and polymer separator were integrated into a single polyester cotton textile layer. After vacuum impregnating the battery with an aqueous electrolyte, the zinc-ion battery device achieved an areal capacity of 35.6 μAh·cm^{-2} between 1.9 V and 0.9 V. The use of the polyester cotton textile as the separator layer holder increased the device thickness, introducing extra encapsulation challenges. Xu et al. [6] reported flexible vanadium (V) oxide carbon-fiber cloth electrodes for potential zinc-ion battery application. This electrode demonstrated an areal capacity of 154 mAh·g^{-1}. These examples demonstrate the capability of fabricated ZIBs within textile material. However, the electrode material vanadium (V) oxide is corrosive and required extra consideration during the device-encapsulation process. It can increase the device thickness and mechanical inflexibility, introducing extra encapsulation challenges for real-world power source units in e-textile systems.

This paper builds upon the previous works [5], the proposed zinc-ion battery was fabricated on top of a polyester-cotton textile using solution-based processes and inexpensive materials. The battery's anode was a flexible zinc polymer film prepared via spray coating, and the battery's cathode was a spray-coated manganese (II, III)-oxide (Mn_3O_4) polymer layer on a polyester-cotton textile. The separator layers were implemented on top of the battery's cathode with doctor blading or screen printing followed by a phase inversion process. The battery was tested with an aqueous electrolyte to study its discharge performance.

2. Material and Methods

Figure 1 shows the fabrication process of the zinc-ion battery. Firstly, silver paste (TC-C4001, Smart Fabric Inks Ltd., Southampton, UK) was screen printed on the hotmelt PEVA encapsulation film that dried at room temperature for 24 h and polyester-cotton textile where the silver paste soaked through the textile, forming the current collector after cured in a box oven at 100 °C for 30 min. A thin film of nickel metal was sputter coated on top of silver textile forming an inert layer. A Mn_3O_4/polymer (85:15 by wt.%) cathode was spray coated onto the top of the nickel film. The coated textile was then dried in a box oven at 120 °C for 30 min. The current collector and cathode layer were screen-printed with a co-polymer solution and treated with a phase-inversion process to form a porous membrane published previously [5]. This co-polymer membrane acts as the separator of the supercapacitor that prevents electrical short circuitry but allows ions to transfer between the cathode and anode electrodes. The anode layer used to evaluate the battery performance was a spray-coated zinc/polymer (95:5 by wt.%) on top of silver that printed on the encapsulation film. The current collector, cathode, separator combined textiles and the zinc metal anode foil shown in Figure 2a were punched into circular shapes with a diameter of 1 cm, added with an aqueous electrolyte containing 1 M $ZnSO_4$ + 0.1 M $MnSO_4$ through the hole before being heat pressed together to form the full device.

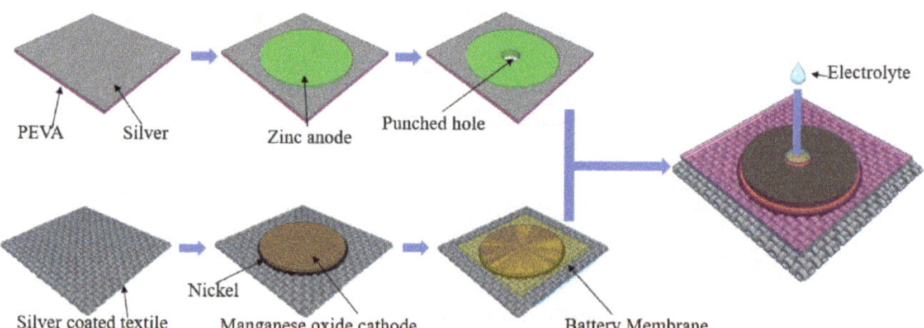

Figure 1. Schematic, illustration of the fabrication process.

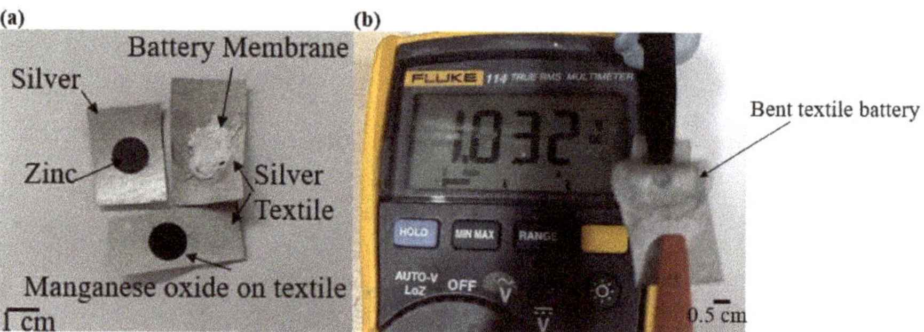

Figure 2. (**a**) Photograph of manganese oxide cathode textile, cathode textile with membrane coating and zinc anode layer with encapsulation film, (**b**) battery initial voltage reading under 90 degrees bending.

3. Results

The manganese-oxide cathode textile, cathode textile with membrane coating and zinc anode layer with encapsulation film are shown in Figure 2a. The device is assessed using a potentiostat Autolab pgsatat101(Metrohm Autolab, Utrecht, The Netherlands). Its electrochemical performance was obtained from galvanostatic cycling (GC) at 1 mA·cm^{-2} and cyclic voltammetry (CV) at 10 mV·s^{-1} between 0.9 and 1.9 V.

The cycling test in Figure 3a was derived from the GC test at 1; this battery achieved an area capacity of 19.1 µAh·cm^{-2} between 1.9 and 0.9 V after the initial test cycle. Figure 3b shows the CV test results after the first test cycle for the zinc-ion battery on the textile. The oxidation (charging) peak occurred at 1.72 V; it also demonstrated a reduction (discharging) peak at 1.26 V. These are typical voltage peaks for the redox reactions in a zinc-ion secondary battery with manganese-oxide cathodes.

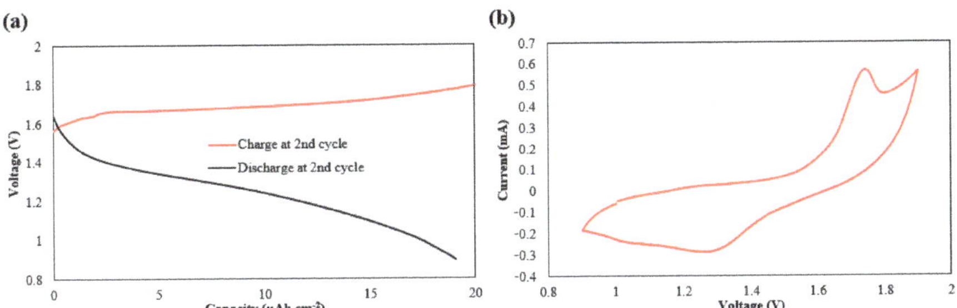

Figure 3. (**a**) GC-derived voltage charge and discharge result of the zinc-ion battery at 1 mA cm^{-2} tested between 0.9 and 1.9 V, (**b**) CV tests between 0.9 V to 1.9 V at the scan rate of 10 mV·s^{-1}.

4. Conclusions

This paper presents an encapsulated, flexible zinc-ion battery on a single piece of polyester cotton. The operating voltage for this textile battery in this work was between 0.9–1.9 V and achieves an area-specific capacity of 19.1 µAh·cm^{-2} and demonstrates good bending durability. In comparison with the previous devices [6], the proposed zinc-ion battery is encapsulated and tested without tube fitting. Future work will include optimizing the formulation and fabrication method of the polymer membrane in the textiles for better electrochemical performances and durability. The final device can be applied in a wide range in the e-textile system.

Author Contributions: Conceptualization, S.Y.; methodology, S.Y. and N.H.; validation, S.Y. and N.H.; formal analysis, S.Y.; investigation, S.Y.; resources, S.B.; data curation, S.Y.; writing—original draft preparation, S.Y.; writing—review and editing, N.H. and S.B.; visualization, S.Y.; supervision, S.B.; project administration, S.B.; funding acquisition, S.B. All authors have read and agreed to the published version of the manuscript.

Funding: The authors thank the EPSRC for supporting this research with grant references EP/P010164/1 and EP/I005323/1.

Institutional Review Board Statement: Not applicable.

Informed Consent Statement: Not applicable.

Data Availability Statement: The data for this paper can be found at DOI: https//doi.org/10.5258/SOTON/D1127.

Acknowledgments: This work was also supported by the Royal Academy of Engineering under the Chairs in Emerging Technologies scheme.

Conflicts of Interest: Steven Beeby is the director of Smart Fabric Inks.

References

1. Zhu, D.; Beeby, S.; Tudor, J.; Harris, N. A credit card-sized self-powered smart sensor node. *Sens. Actuators A Phys.* **2011**, *169*, 317–325. [CrossRef]
2. Carvalho, J.G.; da Silva, J.M. A transceiver for E-textile body-area-networks. In Proceedings of the 2014 IEEE International Symposium on Medical Measurements and Applications (MeMeA), Lisbon, Portugal, 11–12 June 2014; pp. 1–6. [CrossRef]

3. Beeby, S.; Zhu, D. Energy Harvesting Products and Forecast. *Mater. Sci. Found.* **2011**, *9*, 197–218.
4. Dias, T. *Electronic Textiles Smart Fabrics and Wearable Technology*; Woodhead Publishing: Cambridge, UK, 2015.
5. Yong, S.; Hillier, N.; Beeby, S. Fabrication of a Flexible Aqueous Textile Zinc-Ion Battery in a Single Fabric Layer. *Front. Electron* **2022**, *3*, 866527. [CrossRef]
6. Xu, N.; Yan, C.; He, W.; Xu, L.; Jiang, A.; Zheng, A.; Wu, H.; Chen, M.; Diao, G. Flexible electrode material of V_2O_5 carbon fibre cloth for enhanced zinc ion storage performance in the flexible zinc-ion battery. *J. Power Sources* **2022**, *533*, 231358. [CrossRef]

MDPI
St. Alban-Anlage 66
4052 Basel
Switzerland
Tel. +41 61 683 77 34
Fax +41 61 302 89 18
www.mdpi.com

Engineering Proceedings Editorial Office
E-mail: engproc@mdpi.com
www.mdpi.com/journal/engproc

www.ingramcontent.com/pod-product-compliance
Lightning Source LLC
LaVergne TN
LVHW070633170726
843515LV00004B/244

9783036583624